FORGOTTEN HEROES
OF THE
SOUTH AFRICAN WAR
1899–1902

FLORENCE BREED

Published in Australia by Sid Harta Books & Print Pty Ltd,
ABN: 34632585293
23 Stirling Crescent, Glen Waverley, Victoria 3150 Australia
Telephone: +61 3 9560 9920, Facsimile: +61 3 9545 1742
E-mail: author@sidharta.com.au

First published in Australia 1999
This edition published 2022
Copyright © Florence Breed 2022
Cover design, typesetting: WorkingType (www.workingtype.com.au)

The moral right of Florence Breed to be identified as the Author of this Work has been asserted in accordance with the Copyright, Designs and Patents Act 1988.

All rights reserved. Without limiting the rights under copyright above, no part of this publication may be reproduced, stored in or introduced into, a retrieval system, or transmitted, in any form or by any means without the prior written permission of both the copyright holder and the publisher off this book, nor be otherwise circulated in any form of binding or cover other than that in which it is published and without a similar condition being imposed on any subsequent purchaser.

A Catalogue record for this book is available from the National Library of Australia.

Breed, Florence
Forgotten Heroes of the South African War 1899–1902
ISBN (paperback): 978-0-6484916-4-4
pp396

Review

This book is a gem. It has been lovingly put together by Florence Breed and published to coincide with the Centenary of the beginning of the Great Boer War. The work begins with a remarkable chronological listing of the major Boer War events. The author states that the book "is a testimonial to the endurance, discipline and courage of the men from the Donald district who answered their Empire's Call to Arms". But it is much more than that as the manuscript contains letters, diary extracts, maps, cartoons and excellent photographs which tell in the most vivid and colourful manner of the fighting, marching, riding and terrible defeats suffered by the British and Colonial Forces before the tide was turned and the Treaty of Peace was signed on 31st. May, 1902.

For any student of military history, or for any reader interested in this war which for the first time in history saw the use of trench warfare, observation balloons, armoured trains, guerrilla warfare, landmines and concentration camps, this book is a must.

(Reviewed by Harry Powell)

PREFACE

Sometimes a man has no real choice but to take up the sword to defend his weaker brothers and sisters in places where tyrants hold absolute power; therefore this book is dedicated to the ordinary men of Australia who in 1899 did not shirk the battlefield when there was no alternative pathway to defend democracy — not for themselves, but for others.

I believe that the history of wars must be told so that important lessons may be learned by our younger generations. I do not seek to glorify war by painting gory scenes of bloodshed which may arouse an unhealthy desire for carnage. Rather I would wish to emphasize that politicians and presidents must use the conference table as a means of settling disputes and disagreements. We have learnt, or should have done by now, that war never really solves anything because in wars there are never any winners, only losers.

However, it must be said that until the world can find a peaceful solution to deal with evil men who through their greed and lust for power will always be ready to abuse and slaughter their fellow creatures, good men must be ready to take up the sword on behalf of innocent and defenceless victims.

The soldiers whom you will find within these pages went

to fight for others and they displayed a kind of silent heroism which wins no medals. From the pens of these ordinary Australians comes a plain and honest view of the war as they saw and experienced it; often describing an endless horror of heat, flies, dust, polluted drinking water and food shortages. Each day they faced the danger of a bullet in their backs from hidden snipers; and if they were lucky enough to escape the bullets, enteric fever attacked them.

Although I was born exactly thirty years after the last shot in this South African War was fired, my parents told me about Mafeking night when people shouted themselves hoarse, singing "Soldiers of the Queen". Both my father and grandfather served in the Plymouth Division of the Royal Marine Light Infantry, and during their long careers they trained naval gun crews. Grandfather was in the Boer War, and father in World War 1 — on sea **and** land.

Our Donald men arose to do battle for the Empire, but some were doomed never to return to these sunny shores. They were fated to take their last rest beneath the dusty veldt upon which they had marched countless miles, and fought, and died. IT IS OUR DUTY TO REMEMBER THEM.

FIELD GUN CREW PARADE

Proudly marching through Plymouth on August 5th. 1999, this naval gun crew have just won the contest at Earl's Court, London, for the third time. The Royal Naval competition between gunners was originally inspired by their heroic action to transport guns overland to help in the Relief of Ladysmith (3rd. March, 1900) in the heat of South Africa during the great Boer War. Now, after 87 years, this great Naval tradition has ended due to the British Government's defence cuts. The 12-pounder guns and limbers used in the running contest, each weigh close to a ton — the same weight as a Ford Sierra — and crews have to carry the lot over two 5ft. walls and across a 28ft. wide chasm, without men or gun touching the ground.

MATESHIP

Australia's great fighting tradition was born fifteen years earlier than Gallipoli when over 16,000 volunteers sailed from these shores for South Africa to help Britain in the war against the Dutch Boers. Those volunteers proved to be superb horsemen and crack assault troops; in fact, six Australians won the Victoria Cross in that Great Boer War.

Their action at Elands River, when 500 Australians were surrounded by 3,000 Boers and yet managed to hold out for eleven days until help arrived, has gone down in our history books as the "*finest deed of arms in that war.*"

Once an Australian soldier was asked by an Army chaplain what was the religion of the men at the Front and the soldier replied, "*It is just helping one another.*"

There is a Boer War Memorial in the form of a stone statue, in Sturt Street, Ballarat, which shows a life-size trooper on horseback who has ridden back into the midst of the fight, risking his own life to rescue a fellow soldier whose horse has been shot from under him.

Isn't that a wonderful example of Mateship?

PROFILE

JAMES MUIR experienced some very tough times while serving with that famous regiment, the Scottish Horse. It was a dangerous pastime, chasing and fighting roving rebels across the plains and kopjes of South Africa.

One story about James Muir is sufficient to illustrate his prowess as a Mounted Rifleman; and that story came from Police Constable, William Morrison, who was stationed in Donald from 1924 -29. Morrison was a Boer War veteran who had served with James Muir in the famous Scottish Horse. He said that they were in a charge one day against the Boers when James's horse was hit and both animal and rider crashed into the midst of a seething melee of men and horses. He watched in horror as James disappeared from view and he thought he would never see James alive again. Amazingly, Trooper Muir emerged from the fight, uninjured and untroubled except for the minor inconvenience of extricating himself from under his dead horse.

CHRONOLOGICAL LIST OF MAJOR BOER WAR EVENTS

Date	Event
31st May 1899	Bloemfontein Conference
8th September 1899	Britain sends 10,000 men to Natal
26th September 1899	Major-General Symons moves troops to Dundee
27th September 1899	President Kruger calls up burghers
7th October 1899	Lieutenant-General White lands at Durban
9th October 1899	President Kruger Ultimatum
11th October 1899	Outbreak of War
14th October 1899	Start of Siege of Kimberley
	Start of Siege of Mafeking
20th October 1899	Battle of Talana
21th October 1899	Battle of Elandslaagte
24th October 1899	Battle of Rietfontein
30th October 1899	Battle of Modderspruit
31st October 1899	General Buller lands at Cape Town
2nd November 1899	Start of Siege of Ladysmith
21st November 1899	Battle of Willow Grange
23rd November 1899	Battle of Belmont
25th November 1899	Battle of Graspan
26th November 1899	Battle of Derdepoort

Date	Event
28th November 1899	Battle of Modder River
10th December 1899	Battle of Stormberg
15th December 1899	Battle of Colenso
18th December 1899	Field Marshal Roberts appointed Commander-in-Chief General Kitchener appointed Chief of Staff
6th January 1900	Battle of Platrand (Wagon Hill)
10th January 1900	General Roberts lands at Cape Town
24th January 1900	Battle of Spion Kop
5th February 1900	Battle of Vaal Krantz
15th February 1900	Relief of Kimberley
18th February 1900	Battle of Paardeberg
27th February 1900	Surrender of General Cronje
28th February 1900	Relief of Ladysmith
7th March 1900	Battle of Poplar Grove
10th March 1900	Battle of Driefontein
13th March 1900	Capture of Bloemfontein
27th March 1900	Death of Commandant-General Joubert
31st March 1900	Battle of Sannah's Post
4th April 1900	Battle of Reddersburg
14th May 1900	Battle of Biggarsberg
17th May 1900	Relief of Mafeking
28th May 1900	Annexation of the Orange Free State

Date	Event
31st May 1900	Capture of Johannesburg
	Battle of Lindley
5th June 1900	Capture of Pretoria
7th June 1900	Battle of Roodewal
11th June 1900	Battle of Diamond Hill
12th June 1900	Capture of Volksrust
11th July 1900	Battle of Zilikaat's Nek
31st July 1900	Surrender of Commandant-General Prinsloo
27th August 1900	Battle of Bergendal (Dalmanutha)
6th September 1900	Capture of Lydenburg
19th October 1900	President Kruger sails for Europe
24th October 1900	General Buller sails for England
25th October 1900	Annexation of the Transvaal
6th November 1900	Battle of Bothaville
29th November 1900	General Kitchener appointed Commander-in-Chief
13th December 1900	Battle of Nooitgedacht
31st January 1901	Capture of Modderfontein
10th February 1901	Invasion of Cape Colony
28th February 1901	Middleburg Peace Talks
8th May 1901	High Commissioner Milner sails for England
17th September 1901	Battle of Blood River Poort
26th September 1901	Battle of Fort Itala

Date	Event
11th October 1901	Capture of Captain Scheepers
30th October 1901	Battle of Bakenlaagte
7th November 1901	Colonel Hamilton appointed Chief of Staff
16th December 1901	Capture of General Kritzinger
25th December 1901	Battle of Tweefontein
17th January 1902	Captain Scheepers executed
7th March 1902	Battle of Tweebosch
26th March 1902	Death of Cecil Rhodes
4th April 1902	Siege of Ookiep
11th April 1902	Battle of Rooiwal
12–18th April 1902	Peace Delegation meets in Pretoria
6th May 1902	Zulu attack on Holkrantz
15–18th May 1902	Boer Delegation meets in Vereeniging
31st May 1902	Peace Meeting in Vereeniging
	Peace Treaty signed in Pretoria

BOERS: early settlers in South Africa from Holland (Dutchmen)
VELDT: an open, wild plain in South Africa.
KOPJE: a steep, rocky Mil in the open countryside of South Africa.
DONGA: a steep ravine between mountains.
LAAGER: a camp, usually a circle of wagons.

*November, 1899.
Boer soldiers leave for the front.*

BOER WAR, 1899.

The war in South Africa started when the Boers invaded Natal, a British Colony. Britain sent troops to suppress the rising, but the Boers outnumbered the British and anyway they were fighting on familiar ground. The week known as the "Black Week" saw heavy defeats for the British and the Boers were jubilant and confident of victory. But early in 1900, heavy reinforcements of troops from the British Colonies, such as Australia and Canada, turned the tide of this war and the Boers were defeated in a series of battles. Unfortunately, the war continued until 1902 because the Boers used guerilla tactics and their ringleaders were good at hiding.

Private Carlile, of Drouin, who left with the Bushmen's Contingent, writes as follows to a friend in Drouin: -

18th. **April** — have arrived at Beira at last, but are still on board, although the last of the horses left the boat this morning. There is no pier here so the horses had to be slung overboard into rafts and pulled to the shore by a launch where the niggers assted us to unload them. Some of the horses were very stiff after being cramped up for five weeks without being able to lie down. We lost 2 horses between Melbourne and Capetown, 3 between Cape and Beira.

White men do all the overseeing here, and The wages paid to a capable man is 5 for a start, and if good at handling the niggers a rise is soon forthcoming. The climate is very trying to white people, a fever peculiar to the tropics and caused by the swamps, carrying off a lot; while it is a terrible place for sunstroke. The other day 38 men of an English regiment received sunstroke, so you may imagine it is pretty warm. It is the worst place ever I was in for mosquitoes. We sleep in our clothes, but they bite through everything. It is a great country for game of all sorts — lions, tigers, hyaenas, elephants, rhinoceri, antelopes and all other animals that belong to hot climates. Snakes are very plentiful. There is a narrow gauge railway from Biera To the Portuguese boundary, a distance of Some 60 miles; but from thereon to Salisbury there is a wider and more substantial line. The narrow gauge is a ramshackle affair and the wonder is that the engine manages to keep on the line. The width is 2 feet with very light rails and engines to match. It is run by niggers, the whites doing the supervising. The highest rate of speed is 10 miles an hour, the carriages jumping about like a dray on a corduroy road.

 21st. **April** — We go ashore tomorrow at 6 o'clock to make a start on the first stage of our journey. We are going by rail to Salisbury (300 miles) where we will be join by some artillery forces, about 1000 strong, with 36 12-pounders, and the English Yeomanry, which with the colonial forces will make a very decent force.. It seems we are bound for the relief of Mafeking, and 'ere long, no doubt, will have some fighting, as word has just come that there is a force of Boers about 60 miles away. At Capetown We were given the name of "Bushrangers" and the same name sticks here. Some of the men, having a night off on shore, charged the police with drawn swords, and they fled for their lives, believing the men were in earnest, so they never attempt to interfere with the "Bushrangers" now.

INTRODUCTION

There is a wide-spread myth that the nineteenth century was a golden era of peace and prosperity, but nothing could be further from the truth. Beginning with that famous Battle of Waterloo in 1815 and ending with the Great Boer War in 1899, history books describe continuous conflicts; and all of them, whether large or small, required heroism and self-sacrifice on the part of the ordinary British soldier.

This war against the Boers (1899-1902) was the bloodiest and costliest conflict for Great Britain since King Henry V took his great army to France and they fought that famous bloody battle at Agincourt, on 25th. October, 1415.

This book, FORGOTTEN HEROES, is a testimonial to the endurance, discipline and courage of the men from our Donald district who answered their Empire's call to arms. It contains eye-witness accounts, diaries and photographs of volunteer soldiers. Their writings not only highlight the sufferings and brutality of war, but also reveal the extraordinary courage of ordinary men fighting in a far-flung corner of the British Empire. FORGOTTEN HEROES is therefore a mixture of military and local history.

Bushmen from the Australian Colonies were soon recognised as amongst the finest of Britain's fighting men.

From the obscurity of country life where they earned a living as boundary-riders, drovers, farmers, blacksmiths, shop-assistants or agricultural-labourers, they would achieve fame and admiration for their courage, resilience and ingenuity in times of danger and hardship. Their shooting skills and scouting abilities were admired by their English counterparts; whilst newspaper correspondents used words like 'loose-limbed', 'dashing', 'fire-blooded', 'keen', 'gallant', 'hard', 'rugged' and 'alert' when describing the Australians.

For our soldiers the worst aspect of this war was the monotonous, weary marching which meant they could never rest in permanent camps, especially during the guerrilla-warfare stage of the war. However, our Mounted Riflemen were certainly better off than poor 'Tommy Atkins' whose ceaseless marching was aptly described in Rudyard Kipling's famous poem "BOOTS".

> **CHORUS:** *"We're foot-slog-slog-slog-slogging over Africa, Foot-foot-foot-foot-slogging over Africa; Boots-boots-boots-boots moving up and down again, There's no discharge in the war."*

At the end of 1900, the British Commander-in-Chief Lord Roberts gave high praise to all his soldiers in a farewell speech:

> *"This South African Force has performed a unique service. For months on end, in fierce heat, and in biting cold, and in pouring rain, you have marched and fought without a halt, and bivouacked without*

shelter from the elements, and you frequently have had to continue marching with your clothes in rags and your boots without soles, time being of such great importance that it was impossible for you to remain long enough in one place to refit. When not actually engaged in battle, you have been continually shot at from behind kopjes by an invisible enemy, to whom every inch of the ground was familiar, and who, using these natural forts (provided by the strange geology of this countryside) were able to inflict severe punishment upon you while perfectly safe themselves. You have forced your way through dense jungles and over steep mountains, through and over which you have had to drag and haul guns and ox-waggons. You have covered with incredible speed enormous distances and often on a very short supply of food; and you have endured the sufferings inevitable in war to sick and wounded men far from their base, without a murmur of complaint and even with cheerfulness. You have acted up to the highest standards of patriotism, and shown kindness and humanity towards your enemies."

There are some who argue that the Boer War was little more than a skirmish, thus belittling its significance in the overall picture of Australia's development. There are others who vilified Great Britain by stating that it was her desire for the gold and diamonds of the Transvaal that was the cause of the war. They forget that it was the Boers who actually started the war by invading the British colony of Natal and laying siege to the three British towns of Ladysmith, Kimberley and Mafeking; and that the real cause was the attitude of the Boers towards the Uitlanders (foreigners) who lived and worked in their country.

Some members of President Kruger's government supported the idea of giving these tax-paying, foreign workers their rightful privileges as "equal brothers of the Republic" — yet it was the President himself and his bigoted followers who won the day as champions of exclusivenesss and racial hatred.

Before Great Britain could come to their aid, the Boers had deported refugees in cattle trucks on a journey of 1000 miles out of Boer territory. For twelve hours they were without food or water and when unarmed men left the trucks in search of food for the famished women and children, they were jeered and mocked and whipped back into the trucks. Thus in one stroke, the Boers got rid of nearly 100,000 mouths to feed before the war had officially begun.

Then, after the Boers had annexed British territory, they forced other refugees at a few hours' notice out of their properties and across the newly-defined borders. It must also be stated that the Boers imprisoned non-combatants and destroyed the farms and property of British loyalists. Are these the actions of a blameless nation?

Yet even worse was the Dutchman's barbaric treatment of the African natives whom he regarded as only fit to be his slaves and used as cheap labour. Britain had abolished slavery in 1833 and could not allow such behaviour to continue within her territories. No wonder African natives were keen to fight on the British side as 30,000 of them did!

The Boers were really a civilian army, not paid by their government for fighting. A Boer wore his ordinary farmer's rough suit and supplied his own rifle, horse and saddle.

His only payment was a handshake from President Kruger. The women were expected to supply the food needs of their fighting men. Eventually, the Boers had to steal boots and uniforms from British soldiers because their own clothes were worn out. Of course, dressed in British uniforms the enemy thus had an enormous advantage when approaching British patrol parties and outpost sentries who naturally would not fire upon what appeared to be their own men.

It was in this war that we saw for the first time the use of trench-warfare, guerrilla-warfare, observation-balloons, armoured-trains, refugee-camps, blockhouse-lines, landmines and a scorched-earth policy. All of these would be used again in following wars throughout the twentieth century.

But the most important aspect of the Great Boer War (1899-1902) was the fact that it was the 'FIRST MEDIA WAR'. For the first time there were men at the front with cameras who could put a human face upon war's bloody battles and who could write first-hand accounts for the British newspapers. Movie pictures had first appeared in 1896, so now people could see film-footage of a war being fought hundreds of miles away and yet looked so vivid and real upon their screens. War correspondents and photographers at the front began a new kind of journalism, one that was strong on human interest and therefore would appeal to the masses back home.

Another important lesson that the British learnt from this war, but were rather slow at learning, was that their officers should dress like the ordinary soldier. In the early stages of the war when officers wore scarlet uniforms decorated with shiny badges, the hidden Boer snipers would have competitions to see who could shoot the highest number of officers. Eventually, all British soldiers dressed alike in khaki, a colour which blended with their background of brown earth and did not make the men such conspicuous targets for enemy snipers. This war certainly gave the British "no end of a lesson".

As for our Australian veterans of this war, many of them fought again in the next great conflict of 1914-1918 and again distinguished themselves. Perhaps it should be pointed out that the concept of *MATESHIP* was definitely born in the South African War when men rode back into the midst of shot and shell to rescue their comrades whose horses had been shot from under them.

Many Australians returned home with very weakened constitutions from the effects of enteric fever. Others had wounds described thus: — injury to back and leg — shot through hand — shot through leg — shot through shoulder — shot through top of head — shot through both thighs — shot through both legs — loss of eye. It seemed as if the variations of wounds to a soldier's body were endless.

Our local South African heroes who joined five hundred Victorian returned soldiers to receive their war medals from the Duke of York at Government House on the 9th. May, 1901, were Troopers W. Kemmis, A. G. Hornsby and E. Moyle. Sadly, Peter Falla was still in hospital at the time.

CONTENTS

Review	iv
Preface	v
Mateship	viii
Profile	ix
Boer War, 1899.	xv
Introduction	xviii
Contents	xxiv
Queen Victoria (1819 — 1901)	1
The Man Who Started It	6
Young Man, Don't Go!	12
Magersfontein	16
The First Australian	21
Profile	25
Slaughtered At Pink Hill	29
Trooper William Mcallistair	36
Australia's Valiant Hearts	40
Fighting Foe And Fever	43

CONTENTS

November 20, 1900: For Queen And Country.	59
Profile — Arthur Gilbert Hornsby	65
A Hero Of Spion Kop	70
Lancashire Fusilier William Waterhouse	74
British Troops Climb A Kopje As The Boers Fire Down On Them	78
Trooper Alfred Bawden	79
Profile — Trooper Alfred Bawden	81
Mafeking Is Rescued!	83
Lieutenant Frederick Stebbins	85
A Brave Little Garrison	88
Three V. C. Heroes	104
The Boers Versus The British Empire	106
Trooper Jack Bolding	107
Profile — Trooper Jack Bolding	111
Profile — Trooper Edwin Moyle	123
On Reaching A River	127
About Returned Soldiers	130
A Race For Life	134
Our Boys In South Africa	139
Roderick Mcswain Of The Bushmen's Contingent	141
Profile — Roderick Mcswain	146
The Siege At Elands River — August, 1900.	150
The Siege At Elands River	158

Trooper Walter Coombs	161
Profile Trooper Walter Coombs	164
James E. Meyer — A Soldier Of The Queen	165
Guerrilla Warfare	177
Profile — William Russell Walder	195
The Scottish Horse"	197
Profile — James Muir	201
Life On Board A Troopship	205
Extracts From A Shipboard Diary	207
A Soldier In South Africa	209
Doctor Newman's Story	232
The Fourth Victorian Contingent	238
Kemmis Brothers Drowned	241
Chasing Johnny Boer	242
Profile — L/Cpl. John Richard Hoare	245
Praise For The Colonial Boys	247
Back From The War — Hoare, Morris And Poppleton	251
Private Thomas Morris	256
Trooper Charles Midgley Of Minyip	259
Australians And The North Lancashires	264
Profile — Alexander Farrell	268
Private H. Schoman	271
Profile: Private Niven Neyland (No. 1133)	274

CONTENTS

Profile: James Leslie Neyland	275
Trumpeter Charles Pearson	277
Fighting In A Guerrilla-Style War	280
James, Laurence And Ronald Muir	283
Profile — Boundary Rider Ronald Thomas Muir	283
Donald's Last Hero Returns Home	289
Trooper James Edward Meyer	291
Profile — James Edward Meyer	299
The Fifth Victorian Contingent	302
Farrier-Sergeant Percy King	308
Profile — Percy King	310
"The Worst Thing That Ever Happened To Australia!"	313
Profile — William Albert Kemmis	315
Profile — Private John Hamilton	317
Profile — James Duncan	317
Profile — Duncan James Mclennan	318
Profile — James Edward Holland	318
Profile — Trooper Richard Ackland Merrett	324
A Gentleman Of The Road	326
The Second Battalion Australian Commonwealth Horse	329
Fourth Battalion Commonwealth Horse	338
The Hero's Return	340
The Treaty Of Peace — 31st. May, 1902.	341

Did You Know That....?	343
"I Joined A Contingent"	350
The Great Boer War	353
"The Old Brigade" By Charlie Murray (1961)	358
South African War 1899-02	360
Major General Roberts S. Baden-Powell	362

5th. June, 1900. Lord Roberts enters Pretoria, the occupation of which virtually ended the South African War. Systematic guerrilla warfare continued under Generals De Wet and Botha until 1902.

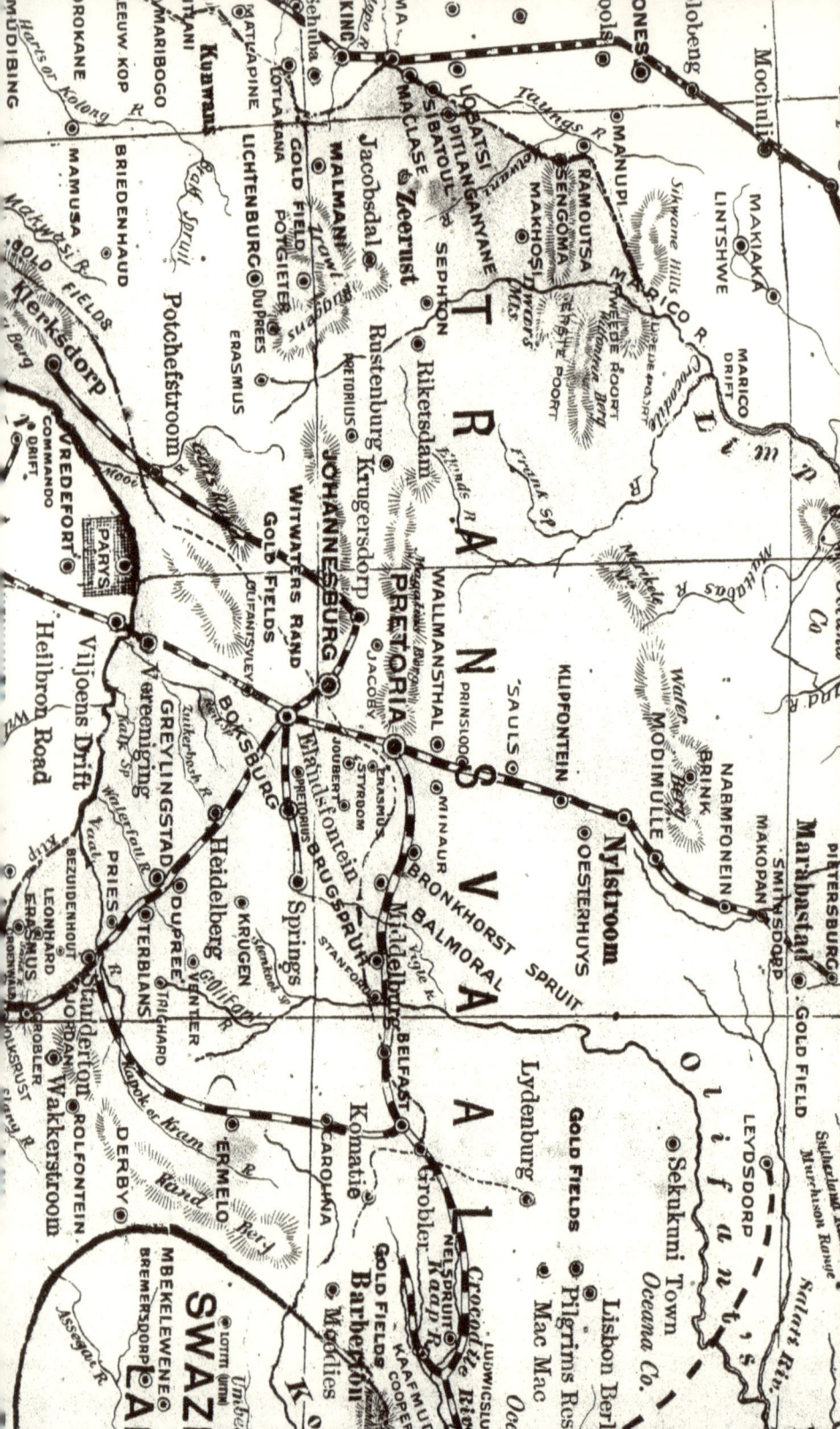

QUEEN VICTORIA (1819 — 1901)

By the year of her death, Queen Victoria had ruled for sixty-four years having been on the throne since 1837. As a girl of 18, she succceeded her uncle, King William IV. Her popularity amongst the ordinary people was truly reflected in the celebrations at her diamond jubilee in 1897 to honour her "Sixty Glorious Years" as their Queen. But it was during the time of the South African War that her kindness and her courageous spirit were most admired; and in appreciation of such virtues her subjects displayed a loyalty and love for her which long outlasted her death.

From the opening of active operations in October, 1899, until her death-bed in January, 1901, this serious conflict occupied her thoughts and determined her actions. The terrible disasters which her soldiers suffered at the beginning of the Boer War caused her great distress and anxiety; but it was said that these British defeats only added fuel to her zeal and she urged her ministers to redouble their efforts in solving the crisis.

When Queen Victoria heard about the slaughter of her soldiers at the Battles of Magersfontein, Stormberg and Colenso, she wept and could not be comforted; yet failure did not quell her spirit and she uplifted those around her by saying, "All will come right!"

The Queen approved British reinforcements for South Africa on an enormous scale at the end of December, 1899, and also approved the leadership of Lord ROBERTS and Lord KITCHENER. When Germany started interfering by sending guns to assist the Boers, Queen Victoria made it clear to the German Emperor (who was her grandson) that she resented his interference and would not tolerate such anti-British behaviour. As a sign of repentence he donated 300 pounds to her fund for the widows and orphans of the 1st. Royal Dragoons.

Near the end of 1899, Queen Victoria bought 100,000 boxes of chocolates to be sent as her personal gift to the soldiers at the Front. On New Year's Day, 1900, the boxes were distributed to all ranks and soldiers were absolutely thrilled with this personal gift from their Queen. Many soldiers sent their unopened boxes home to their families because they felt such a precious gift should be shared and treasured. (These tins are now collector's items for both historical and antique reasons; and some may be found, complete with badly-decomposed chocolates, in military museums amongst the Boer War memorabilia.)

On St. Patrick's Day (2nd. March) in 1900, the Queen gave permission for her Irish troops to wear the Irish national emblem (a sprig of Shamrock) in return for the gallantry they displayed in the South African War; and she spent her vacation in Ireland for the first time in forty years, to meet the people and welcome back the returning Irish regiments.

Throughout 1900 she was busy inspecting the British

Queen Victoria

1819–1901

troops as they left for the war; and she often visited London hospitals, such as the Herbert Hospital at Woolwich Military Barracks, to comfort the wounded soldiers returning from South Africa. With her own hands she knitted woollen comforters and caps for the soldiers — and expressed annoyance when she was told that her handiwork had been distributed only to the officers and none to the privates.

The Queen was genuinely distressed by reports of suffering and consequently would allow no festivities at Court during the war. Some of her personal friends lost their sons in the war. The Queen also had bereavements. On the 29th. of October, 1900, her grandson, Prince Christian Victor of Schleswig-Holstein, fell a victim to this terrible war after he had contracted Enteric Fever on the battlefield. He was the eldest son of Queen Victoria's second daughter, Princess Helena.

Queen Victoria wrote many personal letters of condolences to relatives of officers who lost their lives. No one rejoiced more than she did when the tide of war began to turn in 1900 with the Relief of Kimberley on 15th. February; the Surrender of Cronje at Paardeberg on 27th. February; the Relief of Ladysmith on 28th. February; the Occupation of Bloemfontein on 13th. March; the Relief of Mafeking on 17th. May; and the Occupation of Pretoria on 5th. June. Reports of these favourable events caused great rejoicing amongst her people and in great spirits she warmly congratulated her victorious generals.

The Queen's eighty-first birthday was on Wednesday, 24th. May (1900) so the nation continued its Mafeking celebrations with riotous extravagance and patriotic fervour. The rejoicing during that week was never to be forgotten;

and a **new word** was introduced into the English language to henceforth describe the behaviour of excited large crowds on festive occasions — "MAFFICKING".

Queen Victoria died on 22nd. January, 1901, so the Relief of Mafeking had thus been the last great event of her remarkable reign. The last year or so of her life had been sadly marred by a tragic war and the worry of it all, including the death of a beloved grandson, must have surely hastened her end.

THE MAN WHO STARTED IT

It only takes one man to light a match that starts the lire of racial (or ethnic) hatred and in 1883 that man was Stephanus Johannes Paulus KRUGER, the newly-elected President of the Transvaal, in South Africa. But perhaps a brief portrait of this man may help the reader to understand his motives for going to war against the British Empire in 1899.

He was born of Dutch origin in Cape Province in 1825, and ten years later the Kruger family joined hundreds of other Dutch folks for the Great Trek northwards. They were all moving inland, far from the coast, looking for a district in which they could become free settlers and own the land on which they would build their houses and lay out their farms.

Years earlier the Dutch people had been joined by hundreds of Huguenots driven out of France by religious persecution; and so it was from these two groups of people, the Dutch and French, that the sturdy Boer population of South Africa descended. The word "Boer" simply means "peasant farmer".

In 1835 Paul Kruger's family crossed the Orange River hoping to settle on unoccupied land further north. They travelled across the great veldts (plains) and at night the Krugers slept inside their trek-waggon that was just like a portable house. Of course, they were only one family in a long line of laden waggons and whenever that line stopped all the waggons would immediately assemble in a ring (laager) to make it easier for the people to defend themselves against attacks by natives, or wild animals.

Some of these migrant Boers settled upon land across the Orange River and became the first white settlers of what they called the "Orange Free State", but Paul Kruger's family and others pushed on further north and crossed the Vaal River to found the "Transvaal", a word which means "Across the Vaal".

During this Great Trek, the young Kruger's only source of literature was the Bible which he knew by heart for he had no other formal education. His parents were strict Puritans, belonging to a sect of the Dutch Reformed Church known as "Doppers".

When Kruger was 35 years old he was so religiously fervent that he lay for three days and nights without food or water out upon the lonely veldt, crying aloud to the Lord, begging forgiveness for his sins.

In 1854 Kruger took part in the massacre of Chief Makapan and his tribe of 3000. It was a "revenge" killing for the murder of thirteen Boers. These men had been bartering with Makapan's tribe for ivory and their dishonest dealings had angered the chief so much that he killed them all, flayed their leader alive, and tanned his skin for a shield cover.

Consumed with a passion for revenge, four hundred Boers, including Kruger, surrounded a great cave in which Makapan's tribe was hiding and stood guard both day and night with their guns pointed at the opening. (Similar siege tactics were used by the Boers at Ladysmith, Kimberley and Mafeking.) This siege on Makapan's tribe lasted a week, during which nine hundred natives were shot down as they tried to escape. The rest of the Kaffirs were too afraid to come out of the cave so they eventually became raving lunatics, dying of hunger and thirst.

Then in 1864, Kruger was elected as the leader of the Transvaal "Doppers". Some years later he led a powerful revolt against British administration; and in order to keep the peace Great Britain generously gave the Transvaal a considerable degree of independence. But in 1883 when Kruger was elected the President of the Transvaal Republic, he made it very clear that he wanted **complete** independence from Britain.

Foolishly, he spoiled any hopes of a friendly solution by going behind Britain's back and appealing for military aid from Germany and France. (At first these two countries refused to help his cause, although, later in 1899 at the beginning of the Great Boer War, Germany sent heavy guns and Mauser rifles to help the Boers against the British.)

Kruger's countrymen lovingly called him "Oom Paul" which means "Uncle Paul", and they regarded him with much awe, treating him almost like a god. President Kruger lived in a humble house just like his followers, except that he had two huge marble lions upon his doorstep that had been presented to him as a reminder of the Great Trek in 1836 when Boer trekkers killed 6,000 lions. It was a terrible slaughter and this noble animal never recovered its former numbers.

Kruger had a strange way of talking, just like an Old Testament prophet, spouting prophecies and forecasting terrible retributions for people's sins. He firmly believed that God was on his side in any battle. At the beginning of the war he begged God to direct a Mauser bullet into every British soldier.

The Boers differed greatly from the British over the treatment of African natives. Under Boer rule large numbers of natives had been reduced to slavery, working on Boer farms completely at the mercy of their masters, some of whom treated them very cruelly. Although slavery had been abolished throughout the British Empire in 1833, the Boers still believed an African native was an inferior being and only fit to be the white man's slave. Once when Kruger's black servant did something wrong, he tied the unfortunate Kaffir underneath a waggon making fast his arms and legs to the four axles and then hauled him along in that position for three days. Yet Kruger was known as a very religious man whose conversations were liberally dotted with Biblical quotations.

He certainly possessed outstanding courage. One day while hunting elephants in the jungle, an old rifle exploded in his left hand and the thumb was torn to shreds. Kruger's companions tried to carry him to a doctor, but he refused and instead took out his hunting-knife and quickly amputated the mangled thumb.

There is no doubt that Kruger was a strong-willed man with a wealth of common sense and so it is a great pity that he allowed his hatred of the "Uitlanders" (foreigners) to ruin any chance of peace within the Transvaal. If only he had listened sympathetically to their petitions and grievances, the Transvaal Republic could have prospered with a united population. Instead, racial hatred being much stronger than common sense, it was in vain for any minister to plead for justice on behalf of the Uitlanders. Kruger's feelings towards the newcomers was so bitter that on one occasion he started a public speech with these words, "Burghers, friends, thieves, murderers, newcomers and others!"

When their petitions for the rights of the franchise had been flung back at them by Kruger, these immigrants looked to Britain for help (as many of them were British-born) knowing that their Motherland believed in equal rights and equal duties for all men. *So, will Great Britain stand idly by and watch her subjects (and her authority) in South Africa suffer constant abuse?* As Mr. Chamberlain rightly said, "*No, we must secure for our fellow-subjects in the Transvaal those equal rights and equal privileges which were promised them by President Kruger when the independence of the Transvaal was granted by the Queen, and which is the least that in justice ought to be accorded them.*"

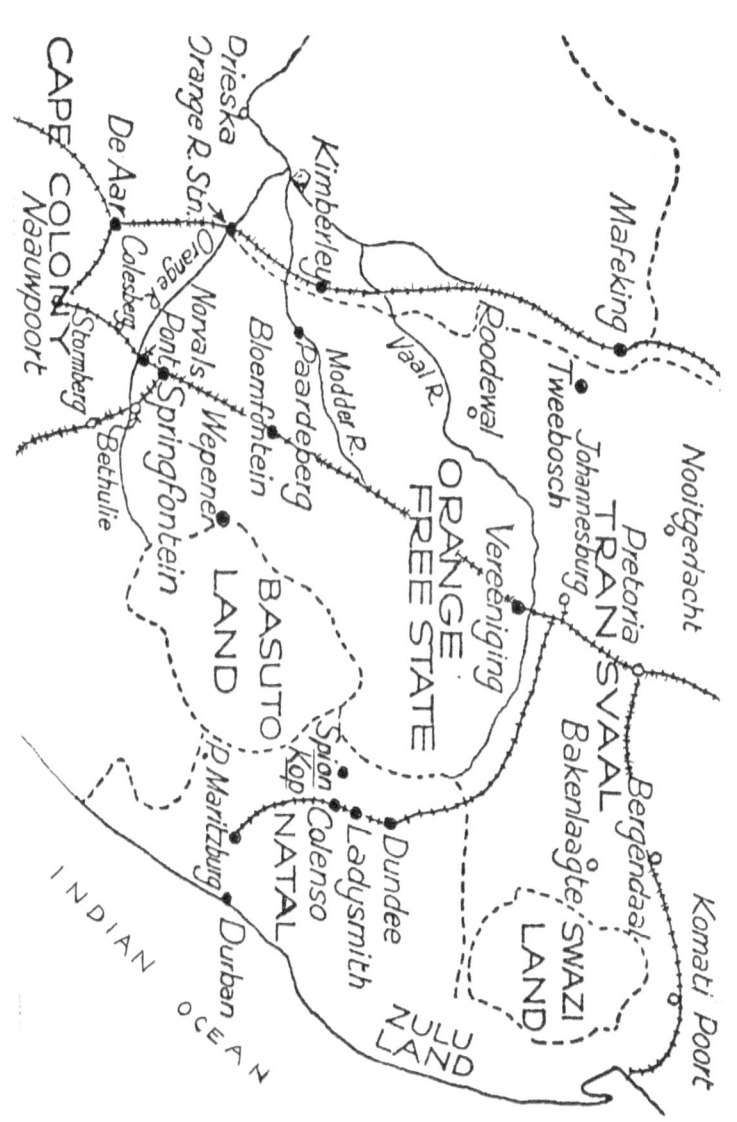

YOUNG MAN, DON'T GO!

In 1914, at the commencement of the Great European War, there was a farmer living in Victoria's far north-west who sent a letter to his young drover urging the young man not to enlist. Based upon his own Boer War experiences, the farmer wrote this following advice: -

Dear Bill, I did intend to have a lot to say about you going to the war, but thinking it over, I have decided that it is none of my business. You have a right to please yourself about what you do. However, I do say think it over well before you take the final step. Remember that when you sign on, you are signing away your liberty and signing your own death warrant.

You will have worse hardships to face than you ever had working for me. You will be treated like a working bullock; you will be loaded up worse than a swagman and have a man over you all day roaring and swearing at you like a bullock-driver; and they don't care if you never get a feed.

Men in camp there will wish they were back here, as it is marching and drudgery all day long on very little tucker. I have seen letters from the first lot to go to the war saying they are having a rough time. Ten hours a day, solid march and drill, with a load on your back and an officer roaring at you, and only a hard biscuit for dinner and no tea.

And after all this hardship, what is next? You are stood up to be shot at; and the first bullet you stop is generally the last thing you will ever see. But then, who cares? You are just one of hundreds shot that day.

And if you live through the bullets, and fever, and dysentery, and come back to Victoria, do the people make you Lord Mayor of Melbourne?

When we came back from the Boer War they gave us a few free dinners and beers. Then after a few weeks they told us to go and get work. That was all the thanks we got, those of us who were lucky enough not to get shot.

You think it over, son.

No doubt when this farmer returned from South Africa he was medically examined before being issued with a civilian suit and discharged; and after that brief business was finished he was left alone to experience the indifference and apathy that follows all wars, as though his nation's gratitude had evaporated with the cheers of farewell at the dockside on his departure to war.

He mentions "BULLETS, FEVER and DYSENTERY", and to the British soldier fighting in South Africa it was almost impossible to escape these ever-present dangers, but add to such terrors the constant agony of sleeping on stony ground at the same time exposed to heavy rains and the bitter cold, and there you have a complete picture of living conditions for our forgotten heroes.

About 17,000 Australians served in this war which turned out to be the bloodiest, costliest and most humiliating war ever fought by the British Army during Queen Victoria's long reign; and so the nineteenth century ended with the British Empire and her colonies catapulted into a tragic war involving thousands of men who had to leave their homes to fight in a foreign land.

When Britain called upon her Colonies for assistance, contingent after contingent of Australians volunteered for service against the Boers and subsequently distinguished themselves, proving to all the world that Australia was a nation to be respected, yet this was the first time our country had sent a large number of her sons away to war.

Many historians have seen fit to condemn the reasons for the Boer War (1899-1902) and vilified Great Britain for not using the "conference table" instead of guns; but few would disagree that the individual soldier involved in that war showed a degree of courage that deserves his nation's respect and reverence.

As for our own Australians, their value and importance progressed to the stage where the bulk of Britain's Mounted Infantry consisted of Colonials; and when many of them served again in the 1914-18 war they were able to bring into that conflict all their valuable experience gathered upon the South African veldt.

MAGERSFONTEIN

There are several stone monuments scattered throughout Scotland as poignant reminders of that country's tragic losses in the Boer War. From the ranks of the Black Watch, the Seaforths, the Argyle and Sutherland Highlanders, hundreds of soldiers laid down their lives in the service of their Queen and Empire,

> *Their dust is in the desert and the deep,*
> *And yet, triumphant o'er the grave their spirits never sleep,*
> *But guard the Freedom which they died to save.*

This moving tribute is carved upon a granite clock-monument in Brora, Sutherland, and refers especially to those Scotsmen slaughtered in the terrible battle of MAGERSFONTEIN; a memorable name in the history of Scotland, for no account of the Boer War battles makes more tragic reading than this massacre of Scottish Highlanders on the 11th. of December, 1899.

The slaughter of the Black Watch makes very painful reading even from this distance of a hundred years. It happened like this: -

Lord Methuen concocted the idea of a night-march

followed by a dawn attack upon the Magersfontein range of hills. However, he omitted to use good scouting to find out the enemy's exact position and this dreadful blunder was dearly-paid for with the blood of a Highland Regiment.

Brigadier Wauchope and his Highlanders were ordered to attack the south end of the Magersfontein Hills which were held by the Boers; so, following the chief's orders they marched all night to arrive in time to surprise the enemy's position at dawn. Unfortunately, thick darkness and heavy torrential rain turned their night-march into an absolute nightmare as they floundered through a never-ending muddy quagmire; yet in spite of these hardships the men proceeded cheerfully on their way, making light of their difficulties even in such awful weather conditions.

Imagine a dense column of almost four thousand men slowly advancing in heavy rain across the muddy veldt towards several hills and see the grisly hand of Death already upon them. Invisible to them, but stretching just in front of them, is a long trench bristling with guns and enemy faces. The Boers are there, lurking inside their elaborately-dug ditches fringed with barbed wire, waiting to deal out death and mutilation to unwary approachers.

Suddenly, the rattling of tins attached to trip wires alert the Boers and the Highlanders are taken completely by surprise. Deafened by a thunderous roar of point-blank fire, blinded by the throbbing flame of a thousand rifles, the Scotsmen drop in their hundreds. Howls of rage and agony split the silence of the early dawn as the more unfortunate ones

become entangled in the enemy's hidden wires and are riddled with bullets.

Brave remnants of Major Wauchope's ill-fated column struggle to reform and advance again, true to their character. One little band actually reaches the enemy trenches, returning with some prisoners and reddened bayonets. Then throughout the following hours, without food and water, the wounded Scotsmen lie beneath a burning sun until the backs of their legs are blistered and raw.

A few fellows painfully hobble forward as the bagpipes summon them to renew their efforts; but no-one can succeed against an enemy so comfortably entrenched as the wily Cronje and his men.

Once again the Boers stopped Lord Methuen from marching to the rescue of the besieged town of Kimberley.

Warracknabeal soldier, Trooper OXENHAM, has this to say about the slaughter in a letter to his mother: -

> "I was told that only one of the Highlanders got away out of over 800 that were slaughtered and to hear his story makes one's blood run cold. They say one of their guides on the night-march was a traitor and a true Boer. You see, the Scots fell right into a trap in the darkness of that terrible night.
>
> The enemy's trenches were narrow and deep and well-camouflaged with grass and scrub, and all fenced off with wire-netting and barbed-wire; so that was what saved the Boers. If only the Highlanders could have got through that wire they would have slaughtered the Boers with their bayonets

Their only survivor described how blood was running knee-deep in places; and how wounded horses were trampling over the dead and dying men lying on the ground in rivers of blood."

MAGERSFONTEIN.
THE TERRIBLE SLAUGHTER OF HIGHLANDERS

THE FIRST AUSTRALIAN

Trooper PETER FALLA was the first soldier from an Australian Regiment to be wounded in the Boer War, so it is indeed a great honour for this little Victorian town of Donald that one of her sons should claim such a distinction while fighting for his country.

At the very first intimations of war against the British Empire, PETER FALLA hastened to volunteer for active service in South Africa and was successful in being selected for the First Victorian Contingent (which consisted of 125 Mounted Rifles and 125 Infantrymen).

Another of Donald's sons, ARTHUR HORNSBY, was also selected and by the end of October these two friends were marching with their regiment through the streets of Melbourne to the cheers of thousands of well-wishers. At the dockside Privates Falla and Hornsby boarded the 'MEDIC' and cheerfully waved as their transport steamed out of the harbour to tumultuous shouts from enthusiastic crowds lining the quayside. The sea-voyage took about four weeks.

After disembarking at Cape Town they soon became embroiled in a great and bloody conflict. Their contingent was destined to fight in many battles in Western Transvaal, from Bloemfontein to Komatipoor, including Pretoria and Johannesburg.

Mounted riflemen were urgently needed to protect the huge British columns of infantrymen and valuable supplies, so before long Private Peter Falla found himself on "Escort Duty" — and on 22nd. January, 1900, he was shot down.

Our national newspapers reported in their headlines:

OUR SOLDIERS

'FIRST BLOOD' DRAWN

From the Australian

Regiment

Mounted Rifleman Shot

That "first blood" was drawn from the veins of Private Peter Falla, No. 108, of the Victorian Mounted Rifles, but let Peter write the story of how his army career was finished for he should know best:

> "When our column halted on Monday night, Corporal Ross and I were sent out on patrol-duty to the front of our long train of men and supplies. Just about 6 o'clock we saw two men approaching the outpost line, so we rode out towards them and when we were about 80 yards from them they shouted something which we did not understand.
>
> We called out in reply, but they suddenly turned

Trooper PETER FALLA

their horses and cantered away. Off we went after them, chasing them through thick, scrubby country, but we had only gone a short distance when we found ourselves in the presence of 40 to 50 armed Boers who were lying in ambush.

They did not ask questions, but opened fire upon us at once. Instantly I felt myself hit in the upper part of the right arm, and a few seconds later I was struck again almost in the same place.

I looked towards my corporal and saw him turning his horse. That was enough for me, and following him I galloped back. But it is strange that Corporal Ross and both horses escaped so well because all the firing-rifles were concentrated upon us.

At first I felt no pain or discomfort, but in a few minutes the arm stiffened and I had to drop my rifle. The firing had alarmed the picket which turned out quickly and we were soon inside our lines. There my mates attended to me and bound up the wound in the arm which was bleeding a good deal. Later on I was attended to by Major M. Williams (Surgeon of Western Australia) who accompanied the column."

Falla was wounded by two bullets on the 22nd. of January, 1900, at Cook's Farm (a place about thirty miles from Belmont) when he rode into an enemy ambush. Peter Falla's wounds were very serious and he was in great danger of losing his right arm which had been penetrated by both bullets.

The first wound was caused by a Mauser bullet and that is a clean wound; but about three inches lower was the second wound inflicted by the terrible Martini-Henry bullet which shattered the humerus into splinters, causing a large area of the flesh to be torn away and making a ghastly opening.

At first Falla was sent to the British army hospital at Orange River, but later he was transported to England. After treatment in Netley Hospital, he returned to Australia in the Orient liner OSTERLEY with his arm still broken.

Brilliant Melbourne surgeon, Dr. Charles Ryan, performed a very complicated operation upon Falla's arm which was a great success and Falla was then able to resume his occupation as a wheelwright and blacksmith.

PROFILE

Trooper PETER FALLA was born in 1876, the second son of Mr. P. Falla and his wife, Margaret Hall, who lived in Donald. Mr. P. Falla had a "Wheelwright and Blacksmith" business, an essential and important profession in the days of 'horse-power' and the young Peter followed in his father's footsteps. When he was old enough, Peter joined the local Victorian Rangers Defence Corps and learnt the necessary skills and drills for military service. Therefore, he volunteered immediately the war began and after careful medical examinations and a thorough test of his horsemanship abilities, he was selected as a member of the First Contingent to go to South Africa — **where he became the first Australian to be wounded in any war.** He received the best medical treatment for his shattered arm in both England and Australia, but owing to his severe illness Trooper Peter Falla was unable to attend the presentation of war medals by the Duke of York in Melbourne on 9th. May, 1901. It was a great disappointment for our wounded hero who certainly deserved to receive Royal acknowledgement for his services to the British Empire. Later, Peter decided to travel around the world and while visiting Guernsey he met his future wife who was a native of that island. In 1903 he returned to South Africa and settled in Umtali, Rhodesia, where he felt there was an opportunity to establish himself as the town's blacksmith and wheelwright. Mr. W. J. McIntosh, also a native of Donald, joined him there and together they founded the firm of "McIntosh and Falla" — waggon builders and wheelwrights — and built up a very successful business on their own merits. Peter and his wife, Eunice, had two daughters, Thelma and Gwen. He was an active and well-known citizen of Umtali and when his partner died he continued running the firm. Peter remained a loyal Australian, joining other ex-patriots to celebrate Australia Day and Anzac Day. Peter Falla died in 1955, aged 79, and the town flag was flown at half-mast, proof of the community's respect and affection for this worthy, hard-working citizen.

Parchment Certificate of discharge of No. 108
Private Peter Falla
(Regiment) 1st Contingent Victoria Mtd Rifles
who was enlisted at Melbourne Australia,
on the 15th October 1899

He is discharged in consequence of his having been found medically unfit for further service

Service 1 year 205 days

(Place) Netley
(Date) 7th May 1901
Signature of Commanding Officer } M B Creagh Col. Ass' Adj Gen'l

Description of the above-named man:—

Age 25 yrs   Height 5 ft 8 3/4 ins
Complexion Fresh   Eyes Hazel
Hair Brown   Trade Wheelwright

Marks or Scars, whether on face or other parts of body { G.S. wound of right upper arm received 22.1.00 at Sunnyside South Africa

N.B.—Any person finding this Certificate is requested to forward it, in an unstamped envelope, to the Under Secretary of State for War, War Office, London.

Should this Parchment be lost or mislaid no duplicate of it can be obtained.

Discharge Certificate Of Peter Falla

STEAMSHIP "MEDIC" leaves Melbourne carrying our soldiers of the First Victorian Contingent.

SLAUGHTERED AT PINK HILL

Although Private THOMAS STOCK was not born in Donald he probably worked on a farm in the district for some time as he knew several Donald folks.

Private T. Stock embarked for South Africa on the 28th. October, 1899, with the First Victorian Contingent. He was in a Mounted Rifles Company which consisted of 125 men and 156 horses and their Commanding Officer was Major George Albert Eddy.

All volunteer recruits had to attend the camp of instruction at Langwarrin prior to being despatched to war, yet these early soldiers did not really need a great deal of training and instruction because they had learnt all the necessary skills and drills by serving in their local defence regiments.

When the S. S. MEDIC arrived at Cape Town all men were sent to Maitland Camp for re-equipping purposes, and it was here that an event of historical importance occurred on the 26th. November, 1899. It was decided that ONE Australian Regiment should be formed by amalgamating troops sent from all the various Colonies and Colonel Hoad was to be in charge of it.

This First Australian Regiment was inspected by the British Commissioner of South Africa, Sir Alfred Milner, who said: — "Colonel Hoad, I am delighted to see the Australians here. They are a fine lot of men and look very fit indeed. Their horses are in excellent condition and I am surprised that you only lost one on so long a sea-journey."

On the 21st. of January, 1900, Private T. Stock with his company of fifty Mounted Rifles suddenly came upon an enemy camp (laager) and he described what happened to them in this letter to a Donald friend: -

> "The first day we were out on the veldt three of our men got shot; and one was shot in the arm in two places — his name is Peter Falla. We had camped at a deserted Boer farmhouse, a few miles from Douglas. Towards sundown, a party of our scouts noticed two groups of Boers riding towards them on each flank; and these Boers were trying to get between them and their camp to cut them off.
>
> Our scouts immediately retired and when the Boers saw that they could not cut them off, they dismounted and commenced firing upon our fellows who also dismounted and answered their fire and must have hit one of them for the Boers quickly departed.
>
> The other party of Boers (about twenty in number) moved around to where our guards were stationed in the scrub and the Boer officer came forward to reconnoitre and came within 200 yards of our sentry who was going to shoot him straight away, but the N.C.O. stopped him.
>
> The rest of our picket rode towards this Boer and he made off, but they chased him. Of course he led them straight towards his men so our chaps were surrounded before they knew it. They turned and retired as quickly

as possible and it was then Falla got two shots in his arm — and it is a wonder they were not all shot.

If the Boers had been able to shoot straight our men would have all been killed for sure. I was in a scouting patrol at the time so when we heard shots and saw our men galloping towards us for all they were worth, we made our way towards them to cover their retreat.

Falla was about 500 yards from me when he fell off his horse. It was a good job for him he was able to stay on its back or he would have been taken prisoner. The Boers vanished when they saw my party advancing."

Private Stock's company arrived by train to Nauuwpoort where they had to guard a large quantities of supplies in that strategically-placed railway station. On their arrival, General Kelly-Kenny said: — "Colonel Hoad, I am delighted to see these Australians and to notice the excellent physique of the men and the fit condition in which you have brought them to this station."

It seems that Private Stock and his mates were to prove their strength and courage sooner than General Kelly-Kenny expected, for on the 12th. of February, at Pink Hill, Coleskop, about a hundred of the Victorian Rifles made a desperate attempt to save the lives of their English comrades (a Wiltshire Regiment) who were isolated upon a kopje and completely surrounded by Boer riflemen.

"FORWARD, AUSTRALIA! NO SURRENDER!" yelled Major Eddy as he gallantly led his small band of Australia's brave sons. It was a glorious but tragic charge for they were hopelessly outnumbered by the enemy, yet with true British pluck and daring they refused to surrender and literally rushed into the jaws of Death. Private Stock and the rest of that little band were caught up in a desperate struggle where no mercy was asked and none was given.

It was a magnificent act of valour by our Victorian Rifles and is best described by a young Boer warrior who was an eyewitness of the event and whose praise is genuine and sincere, coming as it does from one of Britain's enemies.

> *"It was a cruel fight,"* **said the Boer**. *"We had ambushed a lot of the British troops — the Wiltshires, I think — and just as we were firing at them we saw about a hundred Australians come bounding over the rocks in the gully behind us. There were two great big men in front of them cheering them on. We turned around and gave them a volley, but to our amazement they kept coming.*
>
> *They leaped over everything, firing as they charged, not wildly but as men who know the use of a rifle, with that quick, sharp, upward jerk to the shoulder, the rapid sight, and then the shot.* **They were rushing to the rescue of the English**.
>
> *It was splendid, but it was madness. On they came and we lay behind the protection of the boulders and our rifles snapped again at pistol range, but we could not stop those wild men until they charged right into a little basin which was fringed all around its edges by rocks covered with bushes.*
>
> *Our men were lying behind those bushes as thick as locusts, and the Australians were trapped. They were far worse off than their English brothers high up on the kopje whom we had surrounded.*
>
> *In the thickest of the fight snd with his sword flashing*

in the sunlight, the fearless leader of this little company, with a voice like a bull, roared out, "FORWARD, AUSTRALIA! NO SURRENDER!" Those were the last words he ever uttered, for a man on my right put a bullet clean between his eyes and he fell forward dead.

We found out later that his name was Major Eddy of the Victorian Rifles. He was as brave as a lion, but a Mauser bullet will stop even the bravest. Following his example, his men dashed at the rocks like wolves and it was awful to see them; they smashed at our heads with clubbed rifles, or thrust their rifles up against us through the cracks in the rocks and fired at us.

One after another they fell down. Their second big leader (Captain McInerney) went down early, but was not killed. He was shot through the groin, but not dangerously. There was another one, a little man named Lieutenant Roberts — he was shot through the heart — but all the other names I forget.

The Australians refused to throw down their rifles and they fought like furies. If only they had surrendered we would gladly have spared them. One of them climbed right on to the rocky ledge behind which Big Jan Aldrecht was stationed. Just as the Australian got on top of the rock a bullet hit him and he staggered, dropping his rifle; so Big Jan jumped up to catch him before he fell backwards into the ravine below, but the Australian struck out at Jan and hit him in the mouth with his clenched fist before he toppled backwards into the ravine far below and was killed.

We killed and wounded an awful lot of them, but some got away — somehow they fought their way out. I saw a *long row of their dead and wounded laid out on the slope of a nearby farmhouse that evening and I cried to look at them lying so cold and still. They had been so brave and strong in the morning, but by evening they were dead, and we had not hated them, nor they us."*

Private THOMAS STOCK was one of several Victorians killed in that engagement, while most of the others were wounded or taken prisoner. Only a small proportion of his company managed to escape by fighting their way out of the ambush.

Lieutenant H. W. Pendlebury   Lieutenant A. J. N. Tremearne

Captain T. McInerney   Major G. A. Eddy

In his definitive history of "The Great Boer War" Sir Arthur Conan Doyle had this to say about the Victorians (Australians) who so valiantly relieved the deadly pressure upon their trapped English comrades: -

"They proved, once and for all time, that amid the scattered nations of the Empire there is not one with a more fiery courage and a higher sense of martial duty than the men from the great island continent . . . . . . Be it said that throughout the whole British Army there was nothing but the utmost admiration for the dash and spirit of the hard-riding, straight-shooting sons of Australia. In a host which held many brave men there were none braver than they.

TROOPER WILLIAM MCALLISTAIR

William McAllistair was a young shop assistant in the township of Donald when the Great Boer War began in 1899. His chums used to call him "Silly Billy", but it was just a friendly nickname. Like many other young men at that time, he rushed impetuously to enlist in a Mounted Infantry Unit. His letter does not give us much information about his army movements, except that he specifically mentions being at Stormberg with General Gatacre. William was a good horsemen and he boasts that he had to teach the English chaps how to ride. It is a fact that Australia was famous as a nation of rough-riders who had the amazing ability to handle any horse and the English soldiers admired them for it.

> **February, 1900** — "I am now in Cyphergot, South Africa, in the 3rd. Division of the N.S.W. Second Contingent, which General Sir William Gatacre commands, and up to now I have had a good time without any mishaps.
>
> The people here, and also the Imperial troops, treat the Colonials very well.
>
> My company went to Cape Town in the S.S. MORAVIA but we were ordered back to East London

where we landed and had a great reception with plenty of refreshments, cigars, cigarettes, etc.

We then marched with some Artillery and Mounted Rifles to Stormberg to assist General Gatacre. But before we arrived the Boers evacuated their positions.

Peter Falla, of Donald, is over here, but I have not come across him yet. Before we started off to Stormberg, the military fellows had a lot of new horses which none of them could ride, so I had to show them.

I can tell you that the countryside around here called the veldt looks the same as the Donald district, flat, dry, and with very few trees."

It would appear that Trooper William McAllistair was sent to Stormberg with the relieving troops and therefore missed seeing General Gatacre's defeat by a cunning and clever ambush on the night of December 9th. (1899). Stormberg was another humiliating lesson for the British which made them look complete fools.

Imagine that a large number of Boers are advancing along the Kimberley railway line into Cape Colony, so General Gatacre's force is sent to drive them back. Gatacre had fewer than 3,000 British troops under him at the time, but he must take great blame for the hasty way he handled the situation. His soldiers were hurried into open railway trucks at three o'clock in the afternoon and there the poor fellows sat for three hours exposed to the burning sun. Eventually they reached their destination at Molteno and started immediately upon a long march without having had enough time for their food and rest.

Arthur Conan Doyle, the British historian, describes how it

was pitch-dark when the tired men began to move out across the gloomy veldt with the wheels of their guns wrapped in animal-skins to muffle the sound. Their destination was only ten miles away, yet for several hours these exhausted men stumbled through wild and rocky country behind General Gatacre until it was only too obvious they had lost their way. A pathetic lack of reliable guides and good scouting (such as the Australians would show them how to do later in the war) was the cause of the disaster that now befell them, for the Boers had been alerted by their spies and were assembled behind rocks high up on the steep line of hills, waiting to surprise and ambush the unwary British.

Suddenly in the early light of dawn, musket-shots broke the silence of the veldt and Gatacre's men upon the plain were utterly exposed to a fierce onslaught. Fortunately, the enemy's aim was not very accurate otherwise many of our soldiers would have died. At that point General Gatacre should have retreated and tried to save his men, but instead he ordered them to rush forward up the kopje which they soon found was impossible to climb. So there they lay, exposed to the enemy's fire from above them. To make matters worse, when the British guns opened fire it proved more deadly to their friends than to their foes because an officer and several men were cut down by their own shrapnel bullets.

As the British began to retreat from the hillside, three powerful Boer guns opened fire and it seemed as if our men would be slaughtered, but the Boers' shells were badly manufactured. Total British losses were no more than twenty-six killed and sixty-eight wounded. During the retreat many of our men dropped asleep by the roadside, utterly fatigued and unable to walk back to Molteno.

TROOPER WILLIAM McALLISTAIR

Infantryman

AUSTRALIA'S VALIANT HEARTS

President Kruger, having heard about our men's fighting qualities, once said, "Ten Boers are equal to one Australian!" Such high praise from an enemy leader must carry great weight.

Below is an official report about just one of their many brave deeds: -

THIS ORDER ISSUED BY MAJOR-GENERAL CLEMENTS

"**Operations at Slingersfontein. 9th. February. 1900**. The General Officer Commanding wishes to place on record his high appreciation of the courage and determination shown by a party of 20 men of the Western Australians under Captain Moor, in the above operations. By their determined stand against 400 Boers they entirely frustrated the enemy's attempt to turn the flank of our position."

In their honour, the position defended by these 20 Western Australians was later named 'WEST AUSTRALIA HILL'. Their story is best told by one of their enemies, a young Boer and eyewitness who speaks without fear or favour.

"*There were about 400 of us,*" **said the Boer**, "*all picked men, and when the Commandant told us to go take the* kopje we sprang up eagerly and dashed down some hills, meaning to cross the gully and charge the kopje where these twenty men were waiting for us. But we did not know the Australians — then. We know them now.

Scarcely had we risen to our feet when they loosed their rifles on us and not a shot was wasted. Every one picked his man and shot to kill. They fired like lightning, too, never wasting lead; and all around us our best and boldest dropped until we dared not face them.

We dropped to cover and tried to pick them off, but they were cool and watchful and threw no chance away. We tried to crawl from rock to rock to hem them in, but they held their fire until our burghers moved, then plugged us with lead until we dared not stir another step ahead, and all the time their British troops were safely retreating, back through the valley where we had hoped to hem them in.

We gnawed our beards and cursed these fellows who played our game as we had thought no other living man could play it. Then once again we tried to rush the hill and again they drove us back; we could not face their fire. To stand upright to cross a dozen yards meant certain death, and many a Boer wife was widowed and a child made fatherless by these silent men above us.

They did not cheer as we came onward; they only clung as close as climbing weeds to the rocks above us and shot as straight as we hope never to see men shoot again.

Our Commandant, admiring those brave few who would not budge before us in spite of being outnumbered, sent an officer to ask them to surrender and promising them all the honours of war, but they sent us word to come and take them. The officer asked three times and at the third time their Sergeant arose and answered, "Aye, God, come any closer and you will find a bayonet in our hands. Go tell your Commandant that Australia's here to stay!"

And they they stayed and fought us hour by hour, holding us back from reaching their column, when but for them the victory would have been ours. We shelled them and tried to rush them, but they only held their posts with stouter hearts and shot straighter when the fire was hottest. We could do nothing but lie there and swear at them and admire them for their stubborn pluck.

They held that hill until their men were safely through; then dashing down the other side, they jumped into their saddles and made off carrying their wounded with them. **They were but 20 men and we were 400."**

FIGHTING FOE AND FEVER

Trooper ARTHUR HORNSBY went with the First Contingent which numbered about 2,000 men of which 543 were Victorians. Their ship "MEDIC" left Port Melbourne on 28th. October, 1899, to rendezvous with other troop ships, including a boatload of New Zealanders, in Albany Harbour, Western Australia. Then they all sailed together in a great convoy for South Africa, arriving there towards the end of November.

Australia's patriotic enthusiasm for the British cause in the Boer War was overwhelming and none could have possessed more entusiasm than ARTHUR HORNSBY and PETER FALLA, the first two men to go from this tiny Wimmera town of Donald, in Victoria The average age of the Australian troops in that First Contingent was 24 and they were all volunteers. "But they are not so young," remarked one officer," as to be over-reckless."

Arthur Hornsby (No. 91) soon found himself thrust into the fight as a Mounted Rifleman, but he was luckier than his chum, Peter Falla. Here he tells the story of Peter's patrol party which was ambushed: -

"On reaching Dover several patrols were sent out to look

Trooper ARTHUR HORNSBY

for Boers. One of our patrols was under Corporal Ross and consisted of six men, including Peter. After leaving camp, Peter's patrol spotted two unarmed Boers just two miles to their rear, so they gave chase and when they had gone about a mile they suddenly found themselves surrounded by a large party of armed Boers who had been waiting in ambush. It was a TRAP. Too late our men realised their danger.

As Peter was leading the patrol he was barely 50 yards from the waiting Boers. On realising they were hopelessly outnumbered, our men turned away in disorder and fled. At once the Boers opened fire upon their retreating figures and after a short distance towards camp Peter said to one of his mates, "I think I am hit." At that very moment a second bullet struck him just below the shoulder of his right arm, thus disabling him.

Peter dropped his rifle, but managed to cling on to his horse with the left hand till they were out of the enemy's range. Then his mates helped him to dismount and examined his wounds. They found that the first bullet had passed through his flesh just above the right elbow, whilst the second bullet had inflicted an ugly, gaping wound in his right shoulder.

They brought him back to camp where he received every possible care at the hands of our medical staff and the N.S.W. ambulance. Next day he was taken to the Orange River Hospital where from all accounts he is doing well. He bore his injuries as only a brave soldier could do!

Another man in Peter's party had a lucky escape. While they were retiring his horse stumbled into a

Bear-Ant hole and threw him heavily to the ground, but his horse did not gallop away so he was able to mount it again amidst a hail of bullets and ride off unharmed."

During the following weeks Trooper ARTHUR HORNSBY was at the receiving end of some very serious attacks by an aggressive enemy: -

*"**ARUNDEL — February, 1900.** Just to let you know that I am still well and have managed so far to escape the shells and bullets fired at us by the Boers. We have been in the thick of the fighting for thirteen days so I am just about played out, doing without sleep for the last five nights and kept at it all the time.*

I was out on patrol all last night and out three different times today, and all the time under fire from the Boers' big gun called 'LONG TOM'. One shell landed about 20 yards from me and Captain Salmon, but luckily did not burst. The night before last we beat a hasty retreat from Rensburg and as you probably know from the newspaper lists, we have had a good many casualties, several being killed, several wounded, and many taken prisoners.

Our horses and men are quite done up and Colonel French has today wired for relief, so maybe tomorrow we shall get back to Naauwpoort. In all probability the Second Victorian Contingent will relieve us and they are welcome to it for I have been under fire three times and had my fill.

Several of our men have ben mentioned in despatches and some recommended for the Victoria Cross. I've just had a wash for the first time im four days. Peter Falla is doing splendidly and is much better off where he is (in hospital)."

Trooper Arthur Hornsby was delighted when he was promoted to "batman" for his favourite officer whom he had first met in Donald some years previously. It was 1896 when Captain Robert Westrup from "Daisy Hill Park", Amherst, visited Donald to talk to the Young Men's Association; and it was Arthur Hornsby who gave a vote of thanks to this lecturer at the end of the evening. Sadly, Arthur was with Captain Salmon for only six weeks when the officer died: -

"**18th. March, 1900. Naauwpoort** — *My Dear Father, Once again I take the opportunity to write you a few lines to let you know I am still well and at this place. Since writing to you last I have had the misfortune to lose my master, Captain Salmon, of Talbot.*

I told you in my last letter that he was sick with Enteric Fever; and after lying in bed for three weeks and getting no better the doctors gave him up on Monday last. He lingered on till Friday morning at 4 a.m. when he passed away quietly. He was buried at 6 p.m. the same day in the churchyard here, with full military honours.

There were only two of our regiment present, myself and Sergeant Ahearn. By his death I have lost a good friend and master.

However, although I lost a good job as batman for Captain Salmon, I was not long in finding another. The day after Captain Salmon was buried, Captain Hopkins who is also ill in the hospital here, sent for me and I arranged to act as his servant. So I am alright once more and unless he plays me the same game as the late Captain Salmon and dies, I will probably not see the battle line again."

On the 21st. March, Arthur wrote to his father with the tragic news that Captain Hopkins had just died from the dreaded Enteric Fever: -

"Fever is very prevalent here, there being about 60 cases in the Naauwpoort Hospital. I am taking all kinds of precautions to protect myself. I'm taking quinine pills and fruit salts, and have eaten no meat for six weeks. At present there are only two of us in our tent, me and Sergeant Ahearn, a chemist from Melbourne who is a very nice fellow. The two of us, with a friend "Tommy Atkins" of the Berkshire Regiment, had our photo taken here the other day."

At the beginning of April, 1900, Trooper Arthur Hornsby wrote home, expressing his belief that the war was almost over: -

"Now, concerning the war. It is my opinion that there will be little or no more fighting as the Boers are now completely scattered and their President Kruger and his friend, General Steyn, have gone into hiding. There may be a bit of a skirmish to capture Pretoria, the Transvaal capital, but the worst of the fighting is over and rebels everywhere are giving up their guns and ammunition and returning to their homes. You see, the Relief of Ladysmith on March 3rd. and the surrender of their

great General Cronje at Paardeberg on February 19th. completely upset their little apple-cart."

Arthur's next letter to his father, Councillor J. R. Hornsby, is dated 19th. May, 1900, and shows that the British have advanced as far as **Kroonstad**.

"We have marched this far, but have got no train yet owing to the Boers destroying all the bridges and blowing up the railway tracks. Some of the bridges being over deep and difficult rivers, large deviations will be necessary, hence the delay to get our train here. But we expect the train today. We have been on half-rations since our arrival, and the horses also are on short supplies as everything has to be brought in by waggons instead of on the trains.

Our trip up from Bloemfontein was a bit rough, plenty of fighting but no pitched battles, purely rearguard action all the way. The first day I got among the shells and bullets which were flying around pretty thick, but I came out O.K. after some close shaves. One shell passed over Captain Kendall and myself, there being about a yard between us, and ripped up the earth nearby.

There were also plenty of pom-pom shells flying about and it was marvellous to see them bursting among horses and men, yet scarcely a man got hit. The engagement was fast and furious, but our Victorian Mounted Rifles had no casualties, although we were in front of our convoy the whole time.

The Inniskilling Dragoons had the misfortune to be deceived by the Boers' white flag trick and lost something like 20 men. These brave fellows have suffered heavily *since they arrived here. Our convoy would do your eyesight good to see; it is an endless string of mule and oxen waggons, sometimes three or four abreast, and must run into thousands of waggons. They will be re-stocked with supplies for our next move which I hope will land us safe in Pretoria — 150 miles — in about a fortnight's time, and there I hope our troubles will be over, for we are all getting anxious for the end of this war.*

I had the pleasure of a good look at Lord Roberts who is a small, cheerful man, but a good general as he has proved, and who is liked by every man under him. Lord Kitchener accompanied him. The weather has turned very cold and having to sleep outdoors does not help us."

Trooper Arthur Hornsby modestly described his part in the fighting, but he was actually participating in one of the greatest marches in modern history. Australian Mounted Riflemen were just one section of Lord Roberts' great army of fifty thousand troops advancing upon the Transvaal, slowly moving towards Pretoria, the seat of Kruger's government. It was a great distance of nearly two hundred and fifty miles from Bloemfontein to Pretoria, but the men did not mind because they were glad to leave behind that pest-ridden, evil-smelling city where hundreds of their soldiers had been struck down by enteric fever.

It was May 1st. when Lord Roberts and his great army set forth to conquer Pretoria and put an end to President Kruger and his rotten government.

Lord Roberts' mighty army was a magnificent sight as it set forth, moving upon a front of twenty miles. The British Infantryman always marched excellently, even though he had to cover twenty miles a day under a burning African sun and carrying a weight of forty pounds on his back.

Trooper Hornsby pointed out in his letter, there was never any severe fighting — just a steady advance by the British and a steady retirement by the Boers, with little loss of life on either side. It was at the expense of their boots and not of their lives that the British Infantrymen successfully won their way to Pretoria; and during their retreat the Boers blew up several railway bridges (that spanned wide and deep rivers) with the intention of holding back the pursuers. However, fast and efficient repair work by the Pioneer Regiment and the Royal Engineers kept the railroad open, even though Trooper Hornsby said he had been waiting a week for the train which would carry his regiment northwards.

Lord Roberts occupied Kroonstad on May 12th. and halted there for eight days before his great army resumed their advance. Trooper Hornsby wrote that they were waiting for supplies. "No looting!" said Lord Roberts, which meant his soldiers had to survive solely on their meagre rations of foul water and bully beef. This long march ended on 5th. June when they entered Pretoria in triumph.

The Relief of Ladysmith stirred emotions of people throughout the British Empire as nothing else had done during their life-time. The great rejoicing in this little township of Donald, as in every other corner of the empire, was only excelled later by the subsequent outburst of rejoicing at the Relief of Mafeking on 17th. May. However, Trooper Hornsby's statement that the war was nearly over seems rather premature considering that it continued for another two years.

"We (Victorian Mounted Rifles) crossed the Orange River into the Orange Free State at Norval's Pont last Friday. There are said to be 800 Boers here, but they will have to surrender to us as we are in front of them and General Gatacre is coming up behind them with his 8,000 soldiers, so the Boers are just like the meat in a sandwich.

I am sending home some local newspapers and a parcel of souvenirs, including my most valuable memento of the Boer War, THE QUEEN'S CHOCOLATES. These chocolates were a personal gift from Queen Victoria to every one of her soldiers fighting in South Africa. We received them at the beginning of this year (1900).

The chocolates are sealed in a handsome red, white and blue tin with Her Majesty's Portrait embossed in the lid. This tin only measures 6 inches by 3.75 inches and contained half-a-pound of Rowntree's chocolates divided into 12 pieces. All the chaps here are going around offering money to anyone willing to sell his tin, but I would not sell it for a hundred pounds.

Also among my souvenirs are Mauser cartridges which I found on the field after the Battle of Belmont (23/11/1899) I have also sent a Boer dum-dum bullet which can cause a terrible wound that usually proves fatal to the victim."

By the end of July, 1900, Trooper Arthur Hornsby was suffering from Enteric Fever and had been sent to the convalescent hospital in Johannesburg.

"**Johannesburg. 10th. August, 1900.** *I am now in the Convalescent Depot in the Railway Goods Sheds waiting for a train. We sleep in three long rows on mattresses on* the floor and get bread, jam, cheese, bacon, mutton and soup each day. We have doctors' parade twice a week and they give us passes to leave the shed daily from 2 p.m. to 6 p.m. I like the town, but it is very quiet as all the business places are closed. There are some splendid buildings, but the streets are very dirty and dusty.

I think we shall soon be sent to Cape Town. I often visit the Church of England Soldiers' Home where "Tommy Atkins" can read, write letters and play games. I enjoy a game of Draughts; and we can get a cup of tea and delicious buns at a penny each.

I now feel much better so I volunteered to go back to the front, but the doctors said I was not fit enough. Anyway, the war is likely to collapse at any time."

On the 12th. August, 1900, Trooper Hornsby had a narrow escape on his way from Johannesburg to a military convalescent hospital near Estcourt, Natal. His train had just crossed the Buffalo River when the lines were blown up.

"We had a close shave when coming here to Mooi River, for after our train passed Newcastle the Boers blew up the line and killed men who were travelling in the train coming up from Bloemfontein and whom we had just waved to. It was the same day the Boers attacked Newcastle."

A railway accident is a terrible thing, but when it is also an enemy ambush it must be appalling. The Boers derailed the train with an explosive device laid on the track and then shot soldiers out of their trucks with a heavy fire from their hiding-places. Wild Commandant Sarel Theron and his gang were responsible for many train-wrecking tragedies, but he and some of his gang lost their lives sometime later, probably blew themselves up playing with explosives!

Lieut, G. G. F. Chomley.
Lieut. S. T. Staugton.
Lieut. J. C. Roberts.
Surgeon-Capt. W. F. Hopkins.
Capt. R. W. Salmon
Capt. D. M'Leish.
Lieut. G. F. Thorn.

OFFICERS MOUNTED RIFLES.

British troops attack the Boers

British and colonial troops at Paardeberg where, under Lord Roberts, they surrounded General Cronje and 4,000 Boers. The Boer forces surrendered in February 1900

HOME FROM THE WAR
Arrival at Donald Railway Station of Privates Hornsby & Moyle
November 19th 1900

Battle of the Modder River in December 1899, when Lord Methuen and his colonial troops (foreground) were defeated by Generals De La Rey and Cronje, losing about 1,000 men. The Boers' victory was mainly due to their policy of digging and defending numerous small and shallow trenches

Forgotten Heroes

1. Lord Roberts, Commander-in-chief of the British Army
2. Lord Methuen, badly defeated by the Boers at Magersfontein in 1899
3. Lord Kitchener, who took over command from Lord Roberts in 1900
4. General de Wet, the South African military leader and politician
5. The South African General Botha

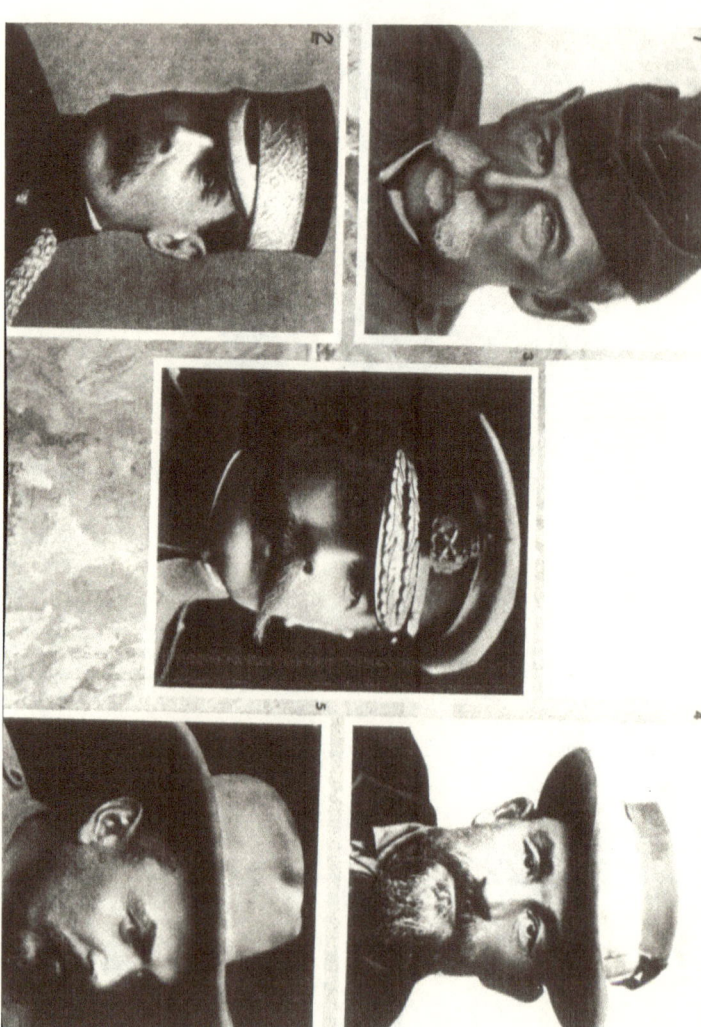

NOVEMBER 20, 1900:
FOR QUEEN AND COUNTRY.

RETURN OF PRIVATES A.G. HORNSBY AND E. MOYLE.

Yesterday afternoon the residents of Donald loyally accorded a most auspicious occasion all the honor it was in their power to bestow. This was the return of Privates **Hornsby** and Moyle from South Africa. Though the keen interest which was some time ago taken in matters connected with the war has subsided to a great extent, this was not apparent in the way in which the residents of Donald treated the return of the two gallant young soldiers, as the town was gay with bunting flying in the breeze--in fact, so great was the demand for flags that the supply ran out at an early hour. From shortly before four o'clock people began to flock to the station, and so great was the crowd that the local detachment of Rangers was called into requisition to keep a small space clear for the reception of the returning soldiers. They were met at the station by their relatives, and by Lieut-Col. Basset and several members of the reception committee, and after exchanging a few hasty greetings with their friends, the pressure of the crowd was so great that some little difficulty was experienced in reaching the drag.

The Donald Brass Band headed the procession, and then

followed the drag in which were seated Privates **Hornsby** and Moyle and their relatives, the Donald detachment of the Victorian Rangers, the Donald Fire Brigade, the Donald Rifle Club, all the Donald State school children, and an immense concourse of people estimated at about 2000. Both of the young men look strong and well, but it is evident that they have not entirely shaken off the effects of the enteric fever which prostrated them, owing to which the medical staff at the front found it necessary to invalid them home. It is hoped, however, that the change to their homes will have the effect of entirely setting them up again, which the voyage from South Africa on board the "Delphic" so happily began. The procession made its way to the Post Office to the strains of the Band. On arriving at that point the procession came to a halt, and Lieutenant Colonel Basset introduced Privates **Hornsby** and Moyle to Cr P.J. Hoban, who was waiting to receive them.

Cr Hoban said: I have today to perform a duty, a very pleasant duty, and that is to accord you, Privates Hornsby and Moyle a very hearty welcome in the name and on behalf of the President, Councillors and ratepayers of the Shire of Donald. (Applause). I am also pleased that God has spared you, not only from the bullets of your enemy, but from the climatic influence of the country which has, unfortunately, proved the grave of so many of your brave brothers in arms. Although there has been disagreements about the war, I am pleased to say that there has been none about the Australian soldiers who have fought for the Empire. The Australians fighting in the South African war have earned high praise from their officers, and from the Commander-in-Chief of the British army, and they have not only established a reputation for courage themselves, but have made the name of Australia respected.. I have again very much pleasure in welcoming you to this your native town and trust that you may not suffer in the future for your experiences in the past (Applause).

Private Hornsby, who was heartily cheered, said he hardly knew how to thank them for the very kindly welcome they had given them that day. They had only tried to do their duty, as every British soldier must do, if called upon. On behalf of Private Moyle and himself he thanked them very much for their hearty reception(Cheers).

Lieutenant-Colonel Basset said he had very much pleasure on behalf of the Donald Detachment Victorian Rangers and on behalf of the 2nd Battalion, in welcoming Privates Hornsby and Moyle on their return.. (Applause).

The procession was again formed, and the two young soldiers escorted to the residences of their friends.

November 23, 1900: The right royal welcome which was extended to Privates **Hornsby** and Moyle was fittingly wound up by a concert in St. George's Hall in the evening. The hall was crowded to the door in a manner which would have delighted the sordid heart of a theatrical manager, until there was not even standing room. Addresses were delivered by the chairman (Lt-Col. Basset), the Rev. W. Hayward, the Rev. A F. White, Mr Carne and Mr Hearn. A very interesting narrative of his experience in South Africa was given by **Private Hornsby**, detailing all that he had gone through from the time he lei) Donald till his return.

November 20, 1900: FOR QUEEN AND COUNTRY.

*Troopers
Arthur Hornsby and Peter Falla
Melbourne, 1901*

TROOPER C. D. COLLES (on the left)

"Perhaps you have heard that I have been selected for the Imperial Contingent for active service in South Africa and I am going into camp on Thursday morning. Of course there is a remote chance of being chucked out in the ballot if there are more men than required. I got through medical examination on Saturday afternoon and it was very severe, but I turned out to be quite sound except for my teeth of which I will have to have a few out before I go away. The riding test was fairly easy — anyhow for me as I had a splendid horse — and I put up top score for my squad in the shooting test. We went to Williamstown and I got six bulls out of ten shots and quite surprised myself."
(Clive Dana Colles — 28/3/1900)

PROFILE – ARTHUR GILBERT HORNSBY

The only son of Mr. James Robert Hornsby, J.P., who settled in Donald during the year 1878 where he established a successful business as the town's carpenter, builder and undertaker, eventually erecting most of the principal buildings in and around the town — and also burying many of its worthy citizens! Arthur was born in Donald, the son of J. R. Hornsby's first wife, Miss Firman, and eventually worked for his father as an apprentice in the building trade. As a young man he trained with the Victorian Rangers and at the beginning of the Boer War enlisted in the First Victorian Contingent with two of his Donald chums, W. A. Kemmis and P. Falla. They went away to the war as Mounted Riflemen. After fighting in many skirmishes against the Boers, Arthur eventually contracted enteric fever and was very ill. He was invalided home to Australia in November, 1900, and arrived back in Donald the same time as Private Edwin Moyle. In May, 1901, Arthur Hornsby and Edwin Moyle travelled together to Melbourne to receive their War Medals from H.R.H. Duke of York at a special presentation. Arthur went to work in his father's timberyard and re-joined the Donald Victorian Rangers where he gained promotion in November, 1901, to the rank of Quartermaster-Sergeant. About this time, the Victorian Police Force was offering jobs to veteran soldiers, so Arthur was tempted to join. Over four hundred returned soldiers and sailors applied, but only 49 were accepted including Arthur. He went to the Police Training Camp and later was appointed to keep law and order in Cavendish, a country town in Western Victoria; and no doubt any criminals wandering onto Constable Hornsby's patch were easily caught for he had had plenty of experience dealing with wily Boers in South Africa. Arthur served as a Mounted Policeman from 21/2/1903 to 31/5/1911 and then returned to Donald. He married his childhood sweetheart, Edith May Crone, who lived opposite.

Mr. RUSSELL MITCHELL, of Donald, said that when his grandfather caught the dreaded enteric fever, the camp cook (a friend) saved his life. This kind cook wrapped Trooper Hornsby in several blankets to sweat out the fever; made a comfortable bed for him beneath a covered-waggon; and then fed him with 100 per cent pure rum each day until he sweated so much that his uniform shrank! But it was more comfortable and hygienic under the waggon than surrounded by fever-germs in a dirty, overcrowded, insanitary hospital-tent.

November 20, 1900: FOR QUEEN AND COUNTRY.

TROOPER ARTHUR GILBERT HORNSBY
VICTORIAN MILITARY FORCES.

CERTIFICATE OF DISCHARGE
FROM

1st Victorian Contingent for Service in South Africa.

This is to Certify that No. 91 Private A. G. Hornsby was Discharged — at his own request — from the 1st VICTORIAN CONTINGENT FOR SERVICE IN SOUTH AFRICA on the 27th November, 1900. Total Service 1 (one) yr. 43 days. Character — Very Good —

VICTORIA.

POLICE DEPARTMENT,
Chief Commissioner's Office, Melbourne, 9th June, 1911.

Register No. 5071

This is to Certify that Arthur Gilbert Hornsby aged 32 years, height 5 feet 9¾ inches, eyes Dark Grey, hair Brown, complexion Dark, a native of Victoria in the Victorian Police Force from the 21st February, 1903, to the 31st May, 1911, during which time his conduct was good. Discharged on account of Resignation.

Dates of Appointment to the different ranks held in the Police Force.

Constable 21.2.03
Senior Constable
Sergeant (2nd Class)
Sergeant (1st Class)

Chief Commissioner.

*Trooper Edwin James Moyle of Donald
in his Boer War uniform
South Africa, 1900*

Trooper A. G. Hornsby of Donald (seated front left) with four soldiers of the East Lancashire Regiment, all convalescing from enteric fever in Johannesburg, 1900.

A HERO OF SPION KOP

Private WILLIAM WATERHOUSE, of Laen, was so keen to fight for his Queen and Empire that he could not wait to join the Victorian Contingents but set sail for England where he was accepted into the Lancashire Fusiliers, a regiment consisting mainly of men from the north of England. In January, 1900, Private WATERHOUSE found himself under the command of General Redfers Buller and suddenly thrust into two of the worst battles in the war.

With an excellent infantry and a strong artillery, Buller's intention was to advance upon Ladysmith to relieve that besieged garrison. Thus Private WATERHOUSE (in the Lancashire Fusiliers) had to march across ridge after ridge of hills, gradually driving the Boers back to Ladysmith. One of the highest kopjes in that region was named SPION KOP (so called because from its summit the Boer trekkers, in 1835, first caught sight of their promised land of Natal).

March, 1900. *"I am still in the land of the living, but I daresay you have seen in the papers that my regiment has been in a couple of engagements.*

I received my baptism of fire at the "BATTLE OF ACTON HOMES". We started attacking the Boers at six o'clock in the morning. My regiment, the Lancashire Fusiliers, formed the first firing-line with three other regiments following us close behind (the Dublin Fusiliers, the Border Regiment and the South Lancashires).

The Boers were concealed behind high kopjes to the right and left of us, and the position we were ordered to capture was directly in front of us; so you can see how hard that task was as we were completely hemmed in by the enemy. But we proved victorious and captured three of their hills. Yet it was nearly 8 p.m. before we took their third hill, so the battle lasted for 14 hours.

Think of it! We were all those hours under shot and shell, bullets flying about us like hailstones, men falling right and left, either killed or wounded; it was something terrible. How I managed to escape from being hit, God only knows. Our regiment lost 70 men on that day, killed or wounded, and we lost 60 more during the following week while endeavouring to hold that position.

It was a terrible week for us. We would start sniping away every morning at daybreak and then the Boers would set the ball rolling and fire back at us. That's the way it went on for a week; the Boers and us firing at each other, until one night we retired to a place called Springfield, about 8 miles away from the Boers, thank goodness.

Here we stayed for a week under canvas and had a rest; which I think was well-earned when I tell you that we had been three nights on the open veldt without any covering. We wore only thin, khaki, summer clothes and you have no idea how cold it can be on the veldt at night. I could not sleep after 2 a.m. so I used to get up and perform jumping exercises to keep my blood in circulation.

Then Lord Buller started off again and we took part in the

"BATTLE OF SPION KOP" on the 23rd. of January. Once again my regiment was in the front firing-line.

There were eight companies of our Lancashire Fusiliers, with several other North of England regiments, and we had to climb 2,000 feet under cover of the darkness with fixed bayonets. At the top of Spion Kop we crouched behind rocks in the misty rain, and then we were ordered to advance in skirmishing order to find out the enemy's position.

We sprang forward with our bayonets and took the Boers' first trench by surprise, but by the time the sun arose we found ourselves surrounded and the Boers gave away their position because they came out from behind their rocks to send a withering fire down upon us.

Enemy shells and rifle fire swept across us and many of our Lancashire Fusiliers were shot. We lay in shallow trenches, enfiladed by enemy rifle-fire and cannon shells, helpless, for several hours in the red-hot sun, unable to move while their bombardment continued.

We received orders to retire that night; and sadly we left that blood-splattered hill-top with its piles of dead and wounded. I tell you the Boers must have lost heavily, too, in that bombardment at Spion Kop because our naval guns gave them "socks; and we all learnt the new rule of war which is — 'Dig your own trench now, or they'll dig you a grave later'. The worst part is we have now been weeks without bread, so when I receive some again I shall treasure it like gold."

Thus ended General Buller's dismal attempt to crush the Boers, but his frontal attack cost the lives of 40 per cent of the brave infantrymen who had fought their way up those steep sides. Nevertheless, in spite of the enemy's withering bombardment our British Infantrymen refused to retreat although there was no cover for them. Men were wounded and wounded and wounded yet again, and still went on fighting.

By nightfall, General Thorneycroft saw the frightful havoc and realised that it was impossible to withstand the relentless enemy artillery any longer, so he ordered a withdrawal. He realised that there would just be a terrible and needless slaughter if he could not give protection to his 4,000 men lying without cover upon the hill-sides. His famous words were: — "Better six good battalions safely down the hill than a bloody mop-up in the morning."

The Drakensburg Mountains.

LANCASHIRE FUSILIER WILLIAM WATERHOUSE

H e took part in a brave attack on Spion Kop mountain where the Boers were strongly-entrenched and somehow he survived those twelve hours of terrible slaughter. One Irish Brigadier said of the infantry's attack, "A finer bit of skirmishing, a finer bit of climbing, and a finer bit of fighting I have never seen." Nevertheless, by the end of the day thirteen hundred British soldiers lay dead and dying on top of Spion Kop.

CROSSING THE TUGELA RIVER

5th. June, 1900. *"I must give you a brief account of myself and regiment since we landed in South Africa,"* **writes Private Waterhouse.** *"It is four months since I left dear old England, sailing from Liverpool on the 13th. December in the "MAJESTIC", arriving at Durban on the 3rd. January after a pleasant voyage. Most of the army corps from England is being drafted into Natal (on the east coast of South Africa) to assist General Buller's forces.*

Leaving Durban we marched up country as far as Estcourt, camping there for a few days and then marched from there to an army base at Frere.

I shall never forget that march. Mind you, it was not a long march, only eight miles, but it took us from four o'clock in the morning till eight at night before we arrived at our destination. It was the worst day possible for marching for it was raining, Heaven's hard, when we started and did not leave off until we arrived at Frere.

Of course the bullock waggons travelled slowly on account of the thick mud and delayed us even more. What we had to eat that day were two biscuits, until we arrived at Frere when we received a toddy of rum to keep the cold out before we retired to bed on the wet, muddy ground.

The next day we left Frere and marched to Pretorius Farm and from there to Springfield where we camped for a couple of days; then we marched at night-time to the River Tugela and bivouacked there.

I told you in the last letter about my baptism of fire at the "BATTLE OF SPION KOP". Well, after General Buller decided we could not advance to Ladysmith along that route, we had to retire back to Springfield where we took a hard-earned rest for a week and had a decent wash.

CROSSING THE TUGELA – AGAIN

On the morning of February 5th. we sallied forth once more with General Buller to have another try to fight our way to Ladysmith. We knew the besieged garrison was desperately short of food and many of them ill with enteric and typhoid fever so we needed no urging to go on.

Well, we crossed the Tugela once more, proceeding to Potgeiter's Drift and we were fighting there for five days at a small hill named Vaalkranz. Although we gained that hill, the Boers increased the number of their guns and our balloonists came in with the information that the Boers were erecting about twenty more guns and if we forced the position we would most likely lose 3000 men.

So the conclusion of our leaders was that the enemy guns were too many, that the way was too hard, and that we must withdraw once again across that accursed river. Therefore Vaalkranz was abandoned and we retired back to camp at Springfield.

OUR ATTACK ON PIETER'S HILL

Then General Buller decided to make another attack at Colenso. Luckily, my regiment (The Lancashires) and some artillery were left behind at Chieveley (about 15 miles north of Estcourt) to act as a line of defence, so we were very fortunate to escape some severe fighting at Colenso on February 21st. where the British losses were about six hundred.

We did not rejoin our Brigade until the 27th. February, the day of the great attack on Pieter's Hill. I tell you it was our Brigade that put the Boers to flight and opened up the way to Ladysnith. We surged up Pieter's Hill, darting and crouching, our bayonets sparkling in the sun. We (the Lancashire Fusiliers) the Lancasters, the South Lancashires, the York and Lancasters, were all together in a long line as we raced for the summit.

"Remember, men, the eyes of Lancashire are watching you!" cried our gallant officers. Our Maxim guns did good work and when we got to the Boer trenches what a sight we beheld.

In one trench there was only one Boer left alive out of thirty; and in another trench were three women, two killed and one badly wounded. It shows you what a cruel and selfish crew they are. I suppose these were some of their wives and when our fire got too hot for them they galloped off leaving their wives to our mercy.

THE RELIEF OF LADYSMITH

We have had rough work to do, but I am glad to say our troops never lost heart although we have had severe losses, yet we knew we had to beat the Boers to save Ladysmith.

On the 1st. March we marched into Ladysmith and what a difference between the two forces. We, the relieving force, marched into Ladysmith looking healthy but mud-stained and as black as chimney-sweeps; and they, the relieved, were spotlessly clean, but sickly-looking and so thin their clothes were hanging on them like sacks. I never saw such a sight before.

But now the relieved garrison is with our column and they look none the worse after their four months' siege.

IS THE WAR OVER?

I really think the hardest part of the war is over now. We have done no fighting since the relief of Ladysmith. At present we are camped 20 miles north of Ladysmith, near the Biggarsberg Mountains, in the north of Natal. The Boers are in these mountains and throw shells at us now and again, and we return the compliment."

BRITISH TROOPS CLIMB A KOPJE AS THE BOERS FIRE DOWN ON THEM

Boer War Memorial to the Gloucestershire Regiment (near Gloucester cathedral)

TROOPER ALFRED BAWDEN

Another Donald man who went off to the Great Boer War was ALFRED BAWDEN, third son of Mr. and Mrs. L. Bawden of "The Oaks", Dooboobetic. Alfred was 26 years of age and working in Western Australia on the gold fields when war began on the 11th. October, 1899. Five months later he caught a boat to Durban and joined Colonel Thorneycroft's famous Mounted Infantry Regiment which was composed entirely of Colonials and Uitlanders.

Colonel Thorneycroft was a great, red-faced, barrel-shaped man who weighed twenty stone, but he had plenty of energy and he trained his men to be very good scouts. Thorneycroft badly needed experienced horsemen like Arthur Bawden, to replace those of his regiment killed and wounded in the massacre at Spion Kop.

Trooper Alfred Bawden would have heard the tragic story of Spion Kop and how on January 23rd. 1900, Sir Redvers Buller's army crossed the Tugela River to attack the Boers who were blocking his road to the besieged town of Ladysmith. Colonel Thorneycroft led his men up Spion Kop and captured its first ridge, but in the darkness and mist he failed to see the enemy entrenched on a plateau above them; and so the enemy was able to fire down upon his men who were confined to a small area on the side of the kopje. The slaughter was pitiful to look upon, with thirteen hundred British dead and dying, their bodies piled up three deep with plenty of unattached heads, arms and legs.

Although Trooper Bawden arrived in South Africa after the Spion Kop disaster, from March onwards he played a vital role in the war as one of Colonel Thorneycroft's scouts, specially trained in field-intelligence work. Of course, it was a dangerous job and eventually Trooper Bawden was shot down on the 10th. of August, 1900, in the manner described beneath: -

> "A small party of 14 men out on patrol were attacked by a force of about 100 Boers and retired fighting back to camp. In the course of the retirement, a Mauser bullet struck Trooper Bawden between the shoulders, coming out at the breast bone. His companions did not abandon his body until life was extinct, but were then forced to leave him on account of the very superior force of the enemy. Later the body was recovered and brought back to camp and buried with full military honours in the evening. His comrades have erected a cross of freestone over his grave which has been carved by two of his company who are stonemasons by trade. Private Alfred Bawden was one of the best soldiers in the regiment and his loss is deeply regretted by officers and men alike. His loss is all the more deplorable coming as it does at the very end of the war when we had thought that the gravest dangers were over." (From a letter written to Mr. and Mrs. Bawden by Captain E. Molyneaux)

PROFILE – TROOPER ALFRED BAWDEN

Trooper Bawden was born in the Donald district on his parents' farm in 1875. His father, Llewellyn Bawden, had been a gold miner on the Ballarat diggings for several years before moving his family northwards to buy some land. So in 1874 the Bawden family settled at Dooboobetic. Alfred left Dooboobetic to seek his fortune in 1895, the time of the gold rush to Coolgardie. Other Donald folks, such as James E. Meyer, also went to the "Land of Promise" and possibly Alfred had heard the story of the Watchem man who found a nugget worth £40 after being there only three days. It is not known if Alfred found any rich spoils during the few years he lived in Western Australia because when the Great Boer War started he sailed directly from Perth to South Africa and joined a British fighting unit there. Trooper Bawden must have been an excellent horseman to get into Thorneycroft's regiment as a scout. It was a dangerous job, out on patrol, exposed to enemy snipers hidden behind rocks. He died fighting for liberty and justice in the service of his Empire; and according to Captain Molyneaux's letter Alfred was a popular soldier of the finest calibre.

(Alfred's father, Llewellyn Bawden, lived to the great age of 93 years.)

The kind captain who wrote a letter of condolence to Alfred's parents thought the war was nearly over; and that was an understandable mistake considering that President Kruger had run away, abandoning his family and his loyal followers to the mercy of the British whom he was always castigating.

It is well-known that when British troops neared Pretoria, Mr. Kruger sent a great amount of coins and gold bullion to Holland. Clutching his stolen bags of money, Paul Kruger ran like a thief in the night, destined to spend the rest of his life in Holland as a fugitive, far away from the country he had ruined. Besides the actual gold, he took Transvaal's great wealth in securities and real estate; and these he turned into cash and had stored in boxes and trunks which were later delivered to him in Holland. He was able to live out the rest of his life comfortably on this stolen treasure estimated to be about 8 million pounds.

Unfortunately the flight of "Old Kruger", as well as the capture of Pretoria on 5th. June, 1900, did not mean the end of the war because owing to the vast area of South Africa and the difficulty of hunting down commandos of mounted Boers it dragged on for two more weary years.

Historians who find it convenient to blame Great Britain for this cruel war should understand that the only person who did well out of it was President Kruger, now remembered by his people as a traitor and a common thief.

MAFEKING IS RESCUED!

Two hundred and sixteen days,
Almost beyond belief!
Give the starving troops great praise
And those who bring relief!
Cheer boys! Make glad noise!
Let the church bells ring!
Fly the flag and let us brag
About glorious MAFEKING!

Three parts of a year of Hell,
Hunger, thirst and horror!
These the things today they suffer,
And defeat tomorrow?
Not defeat — no dust they eat,
Though hunger's pangs they know.
Britons all, great and small,
Cry, "Gallant MAFEKING!"

The glorious news of Mafeking's relief arrived in London on the evening of May 18th. (1900) and the above cartoon appeared in the following morning's "Western Mail". The siege ended on May 16th. It had lasted for 216 days and reflects immortal glory on General Baden-Powell and his small band of soldiers who guarded the little British garrison and refused to surrender. The Boers' plan to starve and shell the townsfolk and soldiers into submission did not succeed. FREDERICK STEBBINS, an Australian and an officer in the Protectorate Regiment, tells his experiences of the famous siege in the following pages.

LIEUTENANT FREDERICK STEBBINS

Lieutenant FREDERICK STEBBINS was an officer in the Imperial Regiment during the Boer War, but prior to the outbreak of hostilities Frederick Stebbins was living in Johannesburg. In this letter to his folks in Melbourne he talks about the worsening political situation over there: -

THE GATHERING CLOUDS OF WAR

20/6/1899. *"Shortly after I came to Johannesburg, three years ago, the Dutch started building a great fort in this town, almost in the centre of the town, on what we call "Hospital Hill", and the hospital now adjoins this fort. The fort has been slowly growing without anybody taking much notice of it, but now it is finished and equipped with Field and Maxim quick-firing guns, and filled up with young Boers and police. You know it's a bit of a cheek, building a fort in the middle of an English community when it is for no other use than to knock their town about."*

Frederick Stebbins' fears were well-founded, for on 11th. October, 1899, the Boers rode to war against the British;

and Frederick joined the Protectorate Regiment only to find himself in the wrong place at the wrong time.

By some strange quirk of Fate, Frederick Stebbins arrived in Mafeking just before the siege of that small place began in earnest. On the 20th. of October, several thousand Boers under their General Cronje surrounded Mafeking and sent in this message to the besieged little garrison, "Surrender to avoid bloodshed!" To which Colonel Baden-Powell defiantly replied, "When is the bloodshed going to begin?"

It was on the 24th. of October that a savage bombardment began which lasted intermittently for seven months. Frederick Stebbins was one of the little town's brave defenders and after the siege ended he wrote a fascinating account of his harrowing experience in various letters to his parents.

THE RELIEF OF MAFEKING

22/6/1900. *"Well, I am safe and sound as the day I was born. You will know Mafeking is relieved at last, thank God; we were relieved on the 17th. May, but only today (the 28th. of May) are we getting a little more bread and food. For the last three months we have been living on "sewers", "dum-dums" and "medals"; the last two are biscuits made out of our horses' oats, crushed up and sifted —* **and also our horses had to be turned into sausages, of which each man had 3/4 lb. a day.**

Well I can tell you we got pretty weak and thin on that diet and having to do patrol-duty night and day just the same; but still, even on that food I believe we could have hung on longer, although I can assure you that all us Mafeking defenders want some nourishing food now the siege is over. Our systems have completely run down.

Oh, about the "sewers" mentioned; they are the tailings, or siftings, of the oats steeped in water until it has turned sour and then the husks floating to the surface are skimmed off. The remainder is boiled up when it has fermented and then dished out as porridge. It is perfectly bitter and sour, and looks like thin flour water. It took a little while before we could swallow it, but when you are starving you must shut your eyes to it, then it goes down with a little screwing of the mouth to get rid of the sourness. Of course, we had no sugar.

A BRAVE LITTLE GARRISON

When 5000 Boers surrounded Mafeking, the garrison inside our little town consisted only of irregular troops, mainly my Protectorate Regiment of 340 men, with 170 Police and 200 volunteers from the citizenry; then add to this number a Town Guard made up of able-bodied shopkeepers, business men and residents, and you will see that our defenders amounted to just about 1000 men.

The weather during the Siege was all that could be desired, the climate being naturally a splendid one here. Our only objection was the prolonged and continuous rainfall which was without doubt a most unique and uncommon one, making brilliant displays of light and consisting of every kind of missile ranging from the common rifle bullet up to the 94-pounder shell which explodes lovely when in its perfect state.

Occasionally we had unexploded shells, about one in every dozen. They were valuable, worth about £5, and when the eagerly sought-after, non-exploding shell arrived in our midst, there was a rush by everyone to pick up the prize. The lucky man would then proceed with great difficulty to carry it away on his shoulder. So, when he gets home he thinks of emptying it for safety purposes, but in the midst of screwing off the nose-cap of that 94-pounder shell he suddenly finds himself exploring Outer Space to reach the Great Unknown from which no man ever returns. I am sorry to say this happened on several occasions, even after we had warned people about it.

"BIG BEN"

For five or six months we never had less than 2,000 or 3,000 Boers around us with their big 94-pounder gun which we nicknamed "Big Ben". The Boers had brought this enormous gun across from Pretoria especially for our entertainment. It could throw a 94-pound shell; so shells by the thousand and bullets by the bushel used to land into our little town daily (the area of which is only a few acres). Of course all this time we lived underground, read books, smoked, played cards, mended socks, sewed on buttons, washed our clothes, and tried to think up new inventions of war.

When it came to our turn to fight what had **we** got? Our full garrison did not consist of 1,200 men, no more than 800 irregular soldiers and the remainder made up of townsmen and natives. Certainly our rifles were as good as the men who handled them, but our guns were all out-of- date, muzzle-loading weapons discarded by the Government years ago. We dug up an ancient 16-pounder gun which was used nearly a century ago and came off some old British war-ship. It was bought by a German trader for old iron and then sold to the native chief who did not know what to do with it so it was left lying on the ground until almost buried with the dust of the years.

Well, we took it to the railway workshops, cleaned it up, mounted it on two trolley wheels and fixed her sights etc. Then we wheeled her up to the front trenches and banged some balls of iron into the Boers. She made a thundering noise and plenty of smoke, but I'm afraid did no harm to the Boers who must have thought we were mad. We called her "Old Sarah".

Our next big gun was made by the railway engineers in their workshops with everything complete; it was a 12-pounder muzzle loader. She used to make a thundering noise and the Boers wouldn't come near her as harmless as she was. We also had one Hotchkiss quick-firing, and six Maxims, but generally two of these were in hospital for repairs most of the time.

OUR METHODS OF DEFENCE

For all those months the town was absolutely naked and empty, with us few half-starved devils holding a thin line of defence around the outside of our town, a distance of six miles in circumference. Just imagine, only a thousand men spread over this distance in our trenches and small forts. Each fort held from ten to forty riflemen and had bombproof shelters and covered-ways for protection as we moved about. The central fort was connected by telephone to all the others. Now you will understand the stillness and silence that prevailed in the inner circle of town.

We had a system of bells by which each part of the town was warned when a shell was coming to enable the inhabitants to run for shelter.

Except for the few men who looked after our food rations and water it was like being in prison; and as time wore on it only made the men more eager to fight despite their weakness and loose trousers. I wish I could make you understand how our emotions were working up and down these last three months.

According to the news we would receive, our expectations would go up from sad and gloomy to cheerful and excited, and then plunge down again. And so it kept on, our hopes going up and down until we got sick of it and would believe absolutely nothing until we actually would see the relieving troops entering our town.

THREE VICTORIA CROSSES

My Protectorate Regiment can now boast of three V.C's all hard-earned and well-deserving — our famous Captain FitzClarence, a Sergeant, and a Trooper are the lucky men; I say lucky, but no, the Sergeant has lost his left arm and leg, therefore it is not lucky for him, is it?

Only twice did my Protectorate Regiment, or rather a portion of it, leave our own lines to assault the Boer trenches ; and although many losses to us, it was done in such heroic style that it took the Boers two or three days afterwards to realise what had really happened to them.

The first sortie was on October 27, when about a hundred men led by our brave Captain FitzClarence moved out from the town with orders to just use their bayonets. We entered their nearest trench with a rush and many Boers were bayoneted before they could emerge from the cover of their tarpaulins; but the enemy guards beyond this captured trench heard a noise and fired wildly into the darkness, thereby killing their own men as well as ours with their rifle fire. Our losses in this gallant attack were six killed, eleven wounded and two prisoners.

Our second sortie (to try and break the enemy blockade) was made on December 26 when two squadrons of the Protectorate Regiment with some of the Bechuanaland Rifles attacked one of their Boer forts to the north of the town; but the enemy had made the fort impregnable and we would need scaling ladders to get in. We met a fierce onslaught of bullets and shells and 53 out of our 80 men were killed and wounded. Captain FitzClarence was wounded, but Vernon, Sandford and Paton were killed as soon as they reached the enemy's guns.

SPIES IN THE TOWN

Yes, it does seem a most remarkable thing to us now it is over, how we managed to hold out so long considering the trouble and annoyance we had to put up with from the Dutch element within the garrison, yet they eagerly sought our protection as soon as war broke out.

Besides defending our lines around the town we had to supply a guard every night to patrol their Laager which contained about 500 Dutch people (whom we also had to feed and that made a great hole in our rations). This Laager was supposed to contain **all** the women in the town, but the white women refused to live in it and preferred to live in their own backyards in large holes, or underground shelters, and risk the shells.

And what did we get in return for feeding and protecting the Dutch? We simply got their spies reporting on all our doings and sending information regularly to the enemy outside.

PRAISE AND ENCOURAGEMENT

Even the Boers admitted they got a good shaking-up from our night sorties and were amazed at the absolute impudence and daring of our men at attempting such feats. Not only here have these daring deeds occurred, but in many other parts of the campaign throughout the country.

The majority of our officers are as plucky and as good as we could wish for, and some have without doubt made lasting fame and glory for themselves.

Often when we seemed abandoned and forgotten by all, words of praise and encouragement would arrive from the outside world. Once it was a special message from the Queen, and once it was a promise of relief from Lord Roberts.

In my letters I cannot possibly give you all the accounts of our fights, engagements, hairbreadth-escapes and shell-dodging; but I **will** tell you about the Boers' last attack on Mafeking.

COMMANDANT ELOFF ARRIVES.

It was Saturday, May 12th. when the Boers attacked us in the gray light of early dawn. Three hundred of them under the command of Eloff crept around to the native side of the town and penetrated into the Stadt (native quarters) which was at once set on fire by them.

It certainly was not quite daylight when they rushed

our Regiment's barracks and took Colonel Hore and twenty of our officers and men as prisoners.

Fortunately, most of us were not in the barracks at the time and as soon as the shooting began we guessed what had happened. The Boers must have fired very erratic and excitedly. The bullets hit the ground up and buzzed over our heads wholesale. Native women were rushing out of the Stadt in all directions, panic-stricken.

ELOFF – HOISTED WITH HIS OWN PETARD

We were soon out of the enemy's range of fire by keeping behind a row of small cottages, and so we pushed our way around the Stadt and barracks, taking up our positions in daylight when the firing seemed to cease. Other little groups of our men had been working their way around also, and when the cordon was almost completed, we had ringed the Boers around in such a way that they could not escape.

The enemy was now divided into two batches and we had secured our different positions; so at last we felt we had a little time to fill our pipes and have a watchful, but comforting smoke, which same smoke has always been our greatest comforter through all our troubles.

There was no point in rushing their positions as they could not possibly go anywhere. During all this time of waiting, occasionally a pipe would be carefully laid down. What for? Just to take a good shot at a Boer's head which every now and again could be seen bobbing above the stone wall of our fort.

Well, this game went on for an hour or two and we began to feel duced hungry, so we sent two men into the town to get our rations, which they succeeded in doing — and I must say we had a good fighting-feed on this occasion. They actually sent us 2 pounds of horse-sausage with the usual 4 ounces of bread per man. We thought this was all right.

"MERCY! MERCY!"

Shortly afterwards, things began to buzz all through the Stadt. About 100 of my Protectorate Regiment in little groups had started Boer-chasing with fixed bayonets; and this is a serious business when you're dodging around huts and corners playing at hide and seek with bullets and bayonets. I can tell you it's awfully exciting and you never know what second you may have to hand in your ticket to join the heavenly choir.

Well, this sort of thing continued throughout the day. The Boers would hold out until our bayonets were right upon them, when they would immediately fall down on their knees, offering up prayers and pleading for mercy. It's difficult to bayonet a man under such circumstances, but this is what they always do when caught.

Here I must give credit to a native called TE KOKO (a Chief's son) who with his 30 natives rendered us valuable assistance in catching the Boers; but he would have assegaied and killed every one of them if we had allowed it. These natives were thirsting for revenge and pounced on them like tigers, but Captain Marsh jumped in and prevented the slaughter.

TREACHERY OF THE WORST KIND

One cowardly instance of Boer treachery was witnessed during the fight; one of our natives was in the act of shooting a Boer at a distance of six yards when the Boer cried for mercy and begged him not to shoot, holding out his rifle for the native to take. So the native spared him and went to take the rifle from the Boer, but as soon as the native's rifle was lowered and he approached the Boer to disarm him, the Boer immediately shot him through the chest. One of our white troopers on seeing this fired three shots into the Boer as he ran away. Whether he was killed we couldn't say as there were so many other dead Boers lying around.

ELOFF SURRENDERS

Just before dark the Stadt (native quarters) was safely in our hands again and the prisoners marched into town. Our next job was now to take the B.S.A.P (British South African Protectorate) fort where Eloff held our officers as prisoners. We were now about 200 yards away and lining the front of the fort along the edge of the Stadt; and we poured in volley after volley for about 30 minutes which must have had a demoralising effect on Eloff's men because they had no hope while their only path of escape was swept with our rifle fire.

We had ceased firing and were just awaiting orders (expecting to make a bayonet charge) when a voice shouted out for all it was worth that Eloff had surrendered to our men in the Fort. At first we were afraid to trust this voice in the dark, and shouted back with not very choice words, what we thought; but the voice persisted in it asking for an officer to come forward.

Then we recognised the voice of Captain Singleton (Dynamite Dick) as he went over the wall. We, the

Troop, followed him, well-extended, and got across that 200 yards like shadows, entered the Fort to find all the Boers in a most frightened condition with their arms and bandoliers thrown all over the ground.

THE GRAND FINALE

It was now 7 o'clock in the evening and we were at last back in our own Fort and 126 prisoners being marched off by 20 troopers.

I think I was the third man to enter the Fort, behind Lieutenant Feltham and Sergeant Ritcher (an Australian), but it will take me too long to tell you all the sights that met us, the destruction they had done and the effects our rifles had on them. They shrank from our bayonets like ones about to be cast into the fire as they abhor the very sight of them. So Colonel Hore whispered, "Unfix your bayonets, boys, the poor fellows are so frightened they'll faint and then we'll have to carry them away."

BRITISH GOOD MANNERS

When Colonel Baden-Powell arrived he said, "Good-evening, Commandant Eloff, won't you come in and have some dinner?" and all our prisoners (Boers, Hollanders, Germans and Frenchmen) were treated to a good supper.

I tell you it was the strangest 15 hours fighting I had ever seen, from 4 a.m. to 7 p.m. Then we had three hours of hard graft getting things tidied up; and 'What Ho' we found the Boers had smashed open the officers' pantry and wine cupboard. It was civilisation once more gazing us in the face. One bottle of wine so satisfied me that I went and had a good sleep — the sleep of an angel

— with a wounded Boer lying on the bed next to me under my charge and debris all around us.

By 4 a.m. next day we were at it again, clearing away dead horses, putting the wounded in hospital and burying our dead.

I have been most lucky and fortunate all through and stand a good chance of coining home. The only thing I have learnt from all this is that "Dutch courage" is like flat beer, and "British morale" is like sparkling champagne."

Frederick Stebbins survived the Siege of Mafeking without any apparent ill-effects, and as a member of the brave little garrison under siege for 217 days he was now the "new face of human heroism". When the news reached London that Mafeking had been relieved, an orgy of celebrations began that quickly spread like an unquenchable fire through all parts of the British Empire.

In the meantime, back in Mafeking it was work as usual for Frederick Stebbins. In a letter to his parents on the 22nd. of June, 1900, he says: -

"Once again the scene is changed and I am sitting in a corner of the large bond store belonging to Whitley, Walker and Co. now turned into a Military Parcels Depot. Talk about presents! Parcels of goods, liquors, and cases of clothing arrive daily from England and Australia, and you could never imagine the quantity that I am arranging for distribution among the Mafeking Garrison, quite a new occupation from what I have had to do this last eight months. But I can assure you the *people from home have not forgotten us! Alas, could we only have had all these things during the Siege our troubles would have been as nothing.*

Every day, troops are arriving here and the town is quite lively and full of bustle. Thousands and thousands of tons of provisions, biscuits, bully-beef, road-rations, flour, meal, regimental uniforms, fresh horses, mules and transport waggons are also arriving here every day for General Hunter's Division, as Mafeking is now the base for General Hunter's column . . .

I am not sending you any 94-pound shells by post, they might get lost. I have been told there is a **Will Stebbins** *in Colonel Plumer's column and that he is a most conspicuous fighter."*

Frederick searched for his lost brother, but without success for there were thousands of soldiers in South Africa and it was like looking for a needle in a haystack. Perhaps this desire to continue looking for his lost brother was the catalyst for Frederick's re-enlistment and by February, 1901, he had joined the 5th. Victorian Contingent and was sailing on the "ORIENT" as a Lieutenant.

While stationed at Maitland Camp all the officers were invited to dinner at "Rondebusch", the magnificent residence of the Hon. Cecil Rhodes: -

"Cecil Rhodes' house and grounds give an impression of unlimited wealth. It was the most beautiful scene I had ever set my eyes on and to describe it would be almost impossible for me to do. It is situated on the side of the famous Table Mountain and the surrounding natural

scenery is perfect. On this mountain, Mr. Rhodes keeps a Zoological Garden full of South African wild animals, but we saw some kangaroos and emus. To approach the house you ascend several terraces of steps, broad and massive, with huge flower beds on either side, and everywhere are beautifully-kept lawns. The splendid dinner was laid out on long tables with servants to wait on us. It was a distinguished company, but everyone wanted to sit beside Mr. Rudyard Kipling. I was nearly the last to sit down, so a waiter squeezed me in between an Imperial Major and the famous poet himself. Sir Charles Metcalfe and Mrs. Kipling sat opposite. Well, right through dinner I was chatting with Rudyard Kipling. He is a fine fellow, very witty and lively. He knows Australia well. About ten years ago when he visited Melbourne, the Editor of the ARGUS newspaper asked him to write an article on the famous Melbourne Cup Race which he said was an honour to do. During the meal he quoted one or two short poems to me and asked if I had heard them before. I said no, but I think he composed them on the spot while he was talking."

By the end of March, the 5th. Victorian Contingent had reached Sunnyside Camp, Pretoria, and the men were preparing to travel to the north of the Transvaal. From Pienaar River upwards that colony was in the hands of the Boers. At Middelburg, on the 30th. April, 1901, Lieutenant Stebbins wrote: -

"Since we left Pretoria we have never stopped more than one night or one day at any place. We have knocked up our horses and many died from wounds received in action so we are now here in Middelburg getting remounts. During the last fortnight we have captured 50 prisoners and taken about 150 people — men, women and children — away from their farms and burnt the buildings before leaving them. These refugees are now quartered in Middelburg in a large camp set apart for them and guarded.

We have thoroughly cleared about 4 0 miles of the country around here, but as soon as we move on some stray Boers will creep back into the hills again, seeking a hiding place from another column that is working behind us. We captured about 2,000 cattle and 90 waggons.

Other columns under General Blood and Lord Kitchener are working within a few miles of us, scouring the country right through and doing exactly the same as we are doing. The result is that there are over 2,000 women and children lodged in Middelburg and fed by the British Government, as well as about 750 prisoners of war. If we keep this going for about six months surely the country will be cleared of Boers and only a few snipers left to worry us.

But the British Government has got an expensive undertaking with all these families who now have no homes and will need so much feeding and looking after. I think it will be another twelve months' work before the guerrilla fighting is over. They are now like the Bushrangers in Australia, split up into small units all over the countryside, and stealing to survive."

Lieutenant Stebbins wrote on October 20th. (1901) from Lack-Kraal, about 30 miles from Utrecht. He is now part of British reinforcements under Colonel Pulteney and still trying to catch the two enemy leaders, De Wet and Botha.

> "We have taken up a position facing the Pongolo Mountain which is about 5 miles long and about 2,000 feet high. It is very steep and precipitous and covered with dense scrub and boulders. It is beautifully rugged, but a fearfully nasty place to have to scale and capture Boers.
>
> We are one of seven columns which surround this mountain because we have at last found Botha's base and hiding-place. There are thousands of cattle on this mount, also General Botha and his commando whom we have been chasing for the past two months.
>
> We have been shelling this mountain for the last two days, but find it one of the ugliest places we have had to tackle yet. We have been waiting for the weaather to clear up. The clouds and fog have been hanging all around the mountain and drizzling rain.
>
> There must be over 10,000 men here waiting to start, but I think we will have a difficult job to capture Botha. Seven British columns are here with Generals Plumer, Bullock, Kitchener, Hamilton, Pulteney (mine) Garnett and Stewart, so you see it is a massive operation."

Indeed, it was a magnificent effort. Feelings of anticipation and excitement were high as seven British commanders hurried to the scene of the action; and our men tried so hard to surround the Boer leader, but failed once again due to the atrocious misty weather and incessant rain which helped Botha and his force to escape. The enemy passed unseen between the British columns and remained hidden elsewhere in the kloofs and forests of that difficult country.

There were many more skirmishes with the enemy before Frederick was able to return safely to his family in Melbourne where he continued his career as a commercial artist. (His work during the siege of Mafeking was highly-praised by Baden-Powell.) Lieutenant F. STEBBINS was truly a hero of the Boer War.

SOLDIERS OF THE QUEEN

THREE V. C. HEROES

Lieutenant Frederick Stebbins has described how he was one of that brave band from his Protectorate Regiment who made a desperate attack upon an enemy fort at Game Tree Hill, near Mafeking, on the night of the 26th. December, 1899, and how their mission failed because the Boers were waiting for them with well-prepared defences. It was a brave onslaught by his attacking party, but a hopeless situation that ended with heavy losses for the Protectorate Regiment. Stebbins wrote that three of his comrades won the highest award for valour on that night at Game Tree; yet he only named one of them, and that was Captain FitzClarence. However, the other two V. C. heroes were Sergeant H. R. Martineau and Trooper H. E. Ramsden and the three citations read as follow: -

> Serg. H. R. **Martineau** — *for attending to the wounds of a comrade under heavy fire during the fight at Game Tree, near Mafeking, Dec. 26, 1899.*

Trooper H. E. **Ramsden** — *for carrying his comrade 600 yards to a place of safety in the face of a hail of bullets during the action at Game Tree, Dec. 26, 1899.*

Captain C. **FitzClarence** *(Royal Fusiliers) for coolness in rescuing an armoured train near Mafeking on Oct. 14, 1899, and for greatly distinguishing himself in engagements outside Mafeking on Oct. 27 and Dec. 26, 1899.*

**To these, and to all our brave lads, we say,
"Your Empire is proud of you!"**

The Boers versus the British Empire

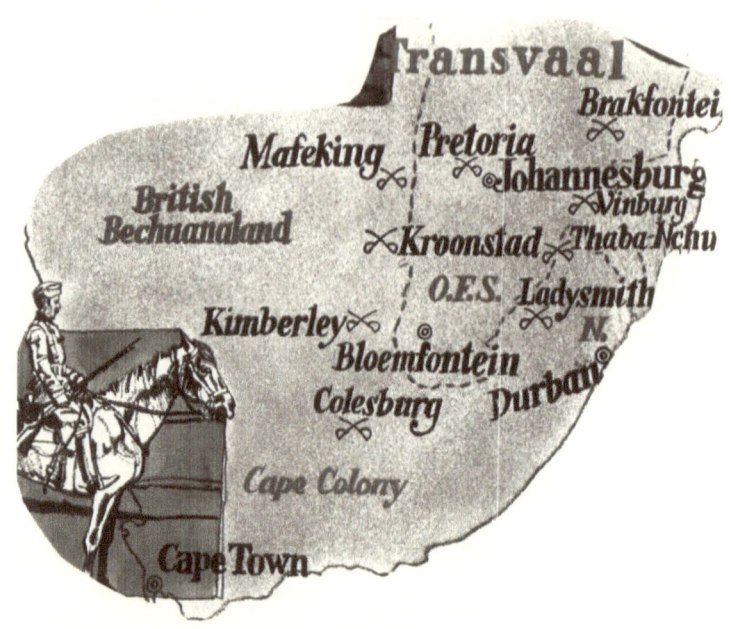

Siege at Elands River

500 Australians surrounded by 2000 Boers for 11 days

TROOPER JACK BOLDING

Trooper PETER FALLA, of Donald, had the honour of being the first Victorian to spill blood in the Boer War, but he was not the only member of his family to go to South Africa because he had two cousins who also volunteered to fight — their names were GEORGE and JACK BOLDING. After spending a period of training at Langwarrin Army Camp, the two Bolding brothers sailed away from Australian shores to take part in that full-scale war.

Trooper JACK BOLDING's diary makes interesting reading, though it is brief as he lost his original diary during the Battle at Bastard's Nek on the 10th. of February, 1900. At the beginning he wrote down a few important Boer words with their meanings, and then he subsequently recorded events in note form:

KOP=HEAD KOOL KOON=TURKEY
WONER=FOWL   ENT=DUCKS

SATURDAY 10TH MARCH. 1900 *We had an easy time today — Lord Roberts had a great victory on Majuba Day (27th. February) with the capture of Cronje — the Boers blew up another bridge across the*

Tugela River — the 2nd. Australian Contingent is on our left — a bit of discontent here over a piece in the Sydney newspapers about us being undisciplined — Joe's birthday — no Boers left in British territory now — Hurrah!

SUNDAY 11 MARCH. 1900 *Arrived at Norval's Pont and now having a day of peace — it's the first Sunday w've been able to call our own and enjoy some quietness and rest — grazing our horses at present — we have been told to learn the Boer words for the phrase, "Come along with us!" which is for the benefit of any Boers we may capture and make our prisoners.*

MONDAY 12 MARCH. 1900 *At Norval's Pont still — we are the hardest-working division in South Africa and nearly all Irish (like the 6th. Dragoons) and hard fighters — we are Major Bucher's picked men — his brother recently died in a reconnaissance balloon accident when actually killed by his own artillery who were working the 40 -pounder guns.*

TUESDAY 13 MARCH. 1900 *Today we rode 15 miles and halted at a Boer farm — had our lunch there — farmer's women were cooking pumpkins — lovely weather now — kopjes are fading away in the distance — bivouacked there.*

MONDAY 26 MARCH. 1900 *We rode out at 5-30 a.m. and crossed eleven miles of a plain covered with grass and stones — only saw a few farms there — it is*

20 miles to the Jagersfontein district — it began raining at 9 p.m. and it poured on us even though we were under cover — our blankets were all wet through — everything is soaked and no firewood to make a fire — I found the following poem on a piece of paper in a prisoner's pocket after we captured a small band of Boers and we all thought it a lot of dreadful cheek: -

"Oh, Tommy, Tommy Atkins, you are the same ruffian of old,
You came to murder our wives and children and sate your thirst for gold;
Why did you not stop on your island and leave us poor farmers alone?
We pity the poor folk of Ireland where you, in the past, have gone."

A BRAVE HORSE

There is a true story about Trooper JACK BOLDING which makes fascinating reading and it proves that soldier's worth. Like all Australian Bushmen, Jack was a fine rider and he had a fine horse that was one of the best in camp. She was a thoroughbred mare, jet-black, with a wild gleam in her eye. Jack boasted that there was no kopje too steep for her to climb when they were out scouring the scrub for Johnny Boer.

One day when Jack was on scout-patrol he observed enemy forces marching in strong numbers towards a gully where he knew twenty-five of his company were camped on observation duty; so he quickly returned to his officer with this vital piece of information.

Unfortunately, there were no other riders available in camp at the time, so Jack received orders to ride out alone and inform his mates of their imminent danger in spite of the fact that Jack's horse was already exhausted after her hard day's work. Still, the officer told Jack he must go and instruct his mates to retire without delay.

Now Jack knew that every minute was precious if he were to save his mates entrenched about three miles away directly in the line of approaching Boers. He wished he could fly like a bird, for between him and their position was a huge, perpendicular kopje covered with great boulders and loose rocks. "A place," thought Jack, "that I wouldn't ask a possum to climb, let alone my horse."

He knew that to take the easier ride all around the base of the huge kopje would mean a waste of precious time. Therefore Jack turned his weary mare's head and in a few minutes she was straining every nerve to gain the craggy summit of that great rock which stood between Jack and the safety of his mates.

His brave mare never faltered in her stride as if she knew men's lives depended on her strength and speed. Horse and rider had almost reached the top of the kopje when bullets from the enemy's Mausers descended upon them like hailstones from the sky. Without a moment's hesitation Jack's brave mare reached the top and began the steep descent down the opposite side.

Bullets spat around the feet of that brave horse as she slid headlong downwards, but Jack's good mare made no mistake and they safely landed at the bottom. When the outpost was reached Jack's mates had no idea that the enemy was almost surrounding them — and if Jack had taken another five minutes to reach them their escape would have been impossible.

Jack's phenomenal ride on his brave steed allowed his

mates sufficient time to get safely back to camp where Jack was praised by officers for his gallant action.

A few weeks' later Jack's brave mare died of blue-tongue fever (rinderpest) and the men whose lives she had saved truly mourned her death, saying, *"She was a beautiful beast and a real marvel; every bit as good as Dick Turpin's legendary Black Bess"*.

PROFILE – TROOPER JACK BOLDING

Bolding returned safely from the war in South Africa, even though he had some very narrow escapes — his horse was shot from under him three times. He became a well-known farmer in the Wonthaggi District of Gippsland, where he married and had six daughters and seven sons, and was elected to serve as a councillor for the Shire of Bass for fifteen years. It seems that Jack became a "Man of Property" because he owned and sold as many as twenty-one farms in the Gippsland area. His descendants say that he never seemed to want to settle down in one place and attributed this restlessness to his youthful days in South Africa when he was chasing Boers across the veldt. Maybe the reason went deeper for Jack was devastated by the loss of his beloved brother, George, who died from Enteric Fever just two weeks after he made a courageous stand during the siege of Eland's River. It is interesting to see that Trooper Jack Bolding's sons, Alan and Murray, continued their father's fight for Freedom when they enlisted in the Second World War (1939-45). However, in 1942 the Japanese army invaded Ambon Island and many Australians were captured, including Alan and Murray Bolding. They suffered terrible privations and beatings for more than three years as prisoners-of-war in the hands of this barbaric, bloodthirsty aggressor. The two brothers bravely endured Japanese cruelty for more than three years, eventually dying from the results of prolonged inhumane treatment in July, 1945, within three days of each other. They must join the ranks of our heroes, too.

OBSERVATION BALLOON

SKETCH MAP OF SOUTH AFRICA TO ILLUSTRATE THE GREAT BOER WAR.

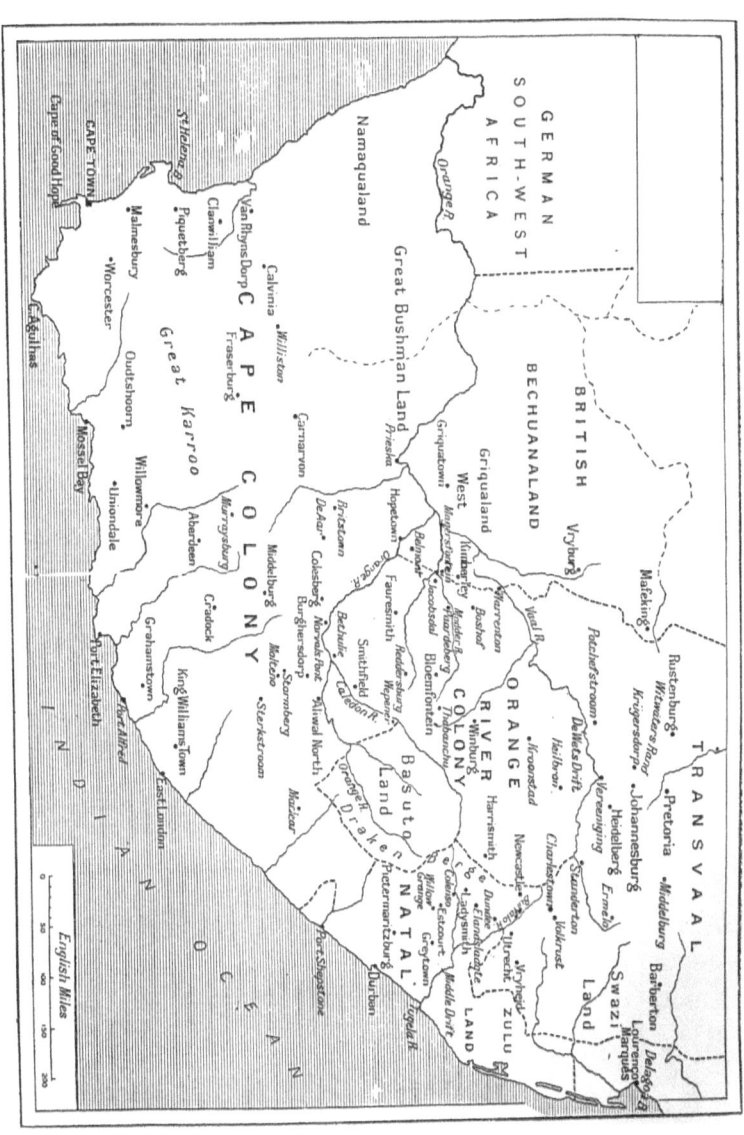

Trooper EDWIN JAMES MOYLE in World War1 uniform

S.S. "Euryalus"

Edwin James Moyle was born on the 9th. August, 1879, at North Laen, the sixth child of Stephen Thomas Moyle and Jane Clark. At the age of 20 he volunteered for service in the South African War; and in December, 1899, of all the young men at that time who volunteered for active service from Donald, only Private E. J. Moyle, was successful. He was the son of Mr. Stephen Thomas Moyle, farmer of Jeffcott South. Edwin sailed with the 2nd. Victorian Contingent on the S.S. EURYALUS which left Melbourne on the 19th. January, 1900, and reached Cape Town nearly three weeks later.

Edwin wrote several letters to his family, describing some of his travels and duties as a colonial soldier: -

> **February 5th. 1900**. — *I am getting on all right now having recovered from sea-sickness. We encountered very heavy weather before reaching Albany and our poor horses were shaking with the wind, cold and wet. I was bad for a couple of days so kept on deck as much as possible and even slept there for two nights. Since leaving Albany we have had a calm voyage and I saw a fully-rigged sailing-ship. She looked very pretty, but was moving rather slowly. We have a lot of work to do on board as each man has to take his turn at piquet duty and keeping guard.*
>
> *It is a rough job looking after the horses as some of them have only a foot's space between their heads and the poor things suffer from sea-sickness. They are so closely stabled that a man can hardly squeeze between them and we are on the go all the time trying to keep their halters and rugs on them. There are 57 horses sick and 2 have died so far. All of them have a job to keep on*

their feet and they get no exercise at all, there being no room to move them about.

I have been "Mess-Orderly" four times and it is a thankless job, running up and down the steps to feed twenty-three men and then washing up their dirty dishes after them. Tell brother Bert (Gilbert Bennetto Moyle) that when he goes to camp he will find out for himself what army life is **really** like!

On this boat we get any amount of tucker with fresh fruit and we have a good time at night, what with concerts and the like, so time passes quickly. Major Wray holds church parade twice on Sundays; and we have morning parade at seven and after that a cigar parade when each man is given two cigars whether he smokes of not.

We had bayonet-practice throughout January and it is a monotonous game, but I suppose it will come in handy when we meet the Boers at close quarters.

We have all had a hair-cut called the "Contingent Clip" which is such a short cut that we look almost bald.

After thirty days of steaming we at last saw the South African coast, a line of scrub-covered hills with pure white sands along the fringe of the blue sea, and no signs of houses or people and it looked so peaceful that you would not think there was a war going on."

S.S. EURYALUS arived at Cape Town on the evening of February 6th. and the men were marched out to Maitland Camp, six miles from Cape Town, where "the sand was up to our knees". After a few days of mounted drill they left for the front.

> "We travelled 700 miles by train, the journey taking two days and three nights. The countryside through which we passed was mountainous and barren, not a blade of grass. The black people we saw along the line cheered us, but the Dutch folks rudely turned their backs on us.
>
> We had to disentrain in pouring rain which delighted the nearby villagers who had seen no rain for twelve months, so they said we had brought it with us.
>
> We will move on to Naauwpoort soon, about 40 miles from here. Our fare is hard biscuits and bully beef for dinner, and for breakfast and supper we have biscuits, tea and coffee. The biscuits are as hard as slate. We passed a Red Cross train on our way up here — it was conveying the wounded down to the Cape."

Naauwpoort must have been a shocking sight as it was just like a "tent city" consisting of a mass of tents and soldiers, mule waggons and duststorms, heat and native mule-drivers. An army of grimy, dusty, khaki-clad soldiers constantly tramped around the place while their horses stood in long rows tied by the head to ropes and by the heels to pegs. Nothing to be seen but rows and rows of tents and rows and rows of horses.

> "On the morning of the 4th. March we were sent out at 4 o'clock on patrol to deliberately draw the Boers' fire so that the Royal Berkshire Artillerymen could see the Boers' smoke and know where to drop their shells. We are not allowed to fire a single shot when scouting although we are armed with Lee-Metford rifles that can fire ten rounds before re — loading.

The enemy was entrenched on the left bank of the Orange River and after six hours of shelling from our men the Boers retired and blew up the Norval's Pont railway bridge behind them. However, that did not trouble us because during the night our engineers built a pontoon bridge across the river and we crossed over next morning into the Orange Free State.

I am in General Clements's brigade (11,000 men) and we are called the "Flying Column". Some of the 1st. Australian Contingent are here with us and they all look well. The Boers are afraid of the Australians as they think we are better shots than "Tommy Atkins".

We are always on the move here, no rest. They say the Boers are getting tired of this war, now they are losing, and are asking for Peace **on their own terms**. What a cheek these Dutchmen have!"

Trooper Edwin Moyle was marching to Bloemfontein, capital of the Orange Free State, in General Clements's force when he was kicked by a horse and consequently sent back to the hospital at Norval's Pont.

"*10th*. **April, 1900**. — *I was in hospital for six days, but I am all right again. Enteric fever is very bad amongst the soldiers and a great number have died from it, including several Victorians. I am back again with my regiment and we are now under Colonel Price. At present the Boers are quiet as far as fighting is concerned.*

We left Orange River on 1st. April for Bloemfontein. It is a ten hours' journey by rail, but a hundred of us rode our horses up and we had a lively time of it when our provisions ran out and we were reduced to one biscuit a day. We are getting accustomed to sleeping out in the rain and on damp and muddy ground, so often we prefer to lie on a bed of stones and keep dry.

We are resting now for a fortnight to recuperate our horses as well as ourselves. Bloemfontein is a pretty little town situated amongst high hills and its streets are planted with shady trees. The inhabitants are Dutch, with a few kaffirs (natives). The Kaffir is a splendidly-made man which is a wonder when one sees the way they are treated when young. It is a common sight to see a Kaffir woman doing her washing in a running creek and a baby on her back with its head bobbing in all directions.

Tell brother Bert to remember me to the members of the Donald Rangers. I am glad to see from the local paper that Donald folks came out so strong at their Patriotic Meeting."

On the 1st. May, 1900, Lord Roberts and his mighty army — which included Trooper Edwin Moyle — left Bloemfontein to march upon Pretoria.

"We have been in action three times since this march began. On the 10th. May we had a hard day of it at Sand River. We started out at 4 a.m. and by sunrise our advance party were into the fight. General Botha and his army had taken up a strong position and were sheltered behind a bank at a bend of the river from whence we could hear the sharp report of their Mausers. My advance party crossed the river and when our rifles spoke you may depend the Boers were not long before they were off. We could see them galloping away as fast as their horses could go!

We did not get far as their artillery was beginning to shell us. The shells came close but did no damage. The ground being very soft and sandy, when they fell they just buried themselves and bursting threw up dirt all around. Our own artillery galloped up amd in half-a-j iff was pounding away in fine style. We then dismounted and blazed away, but the range was too long.

Our guns soon silenced those of the enemy and they retreated. We were then ordered to attack them on their left wing, so away we galloped for three miles and got on top of a ridge and commenced to fire down upon them.

It was thought we were in for a charge with fixed bayonets, but the Boers had had enough and were running away as hard as they could for shelter. They had twelve trains to carry them away, and were soon on their journey to the Vaal River where they intend to make another stand for their country. We only lost one man in that fight. He was Wilkinson of the Australian Horse.

Next day we advanced twenty miles and expected a battle at Kroonstad, but the Boers had left the city unguarded so we had no trouble capturing it on the 12th. May. Unfortunately, we just missed catching President Steyn and his men who got away by the skin of their teeth.

I was the only Donald representative in action that day, so that makes eight engagements I have survived.

Lord Roberts's historic march was drawing to a close. His troops steadily advanced towards their goal while the Boers, tiring over their shoulders, steadily retreated. On May 30th. British troops reached Johannesburg expecting a battle, but General Botha and his army retired without a struggle, abandoning the treasures of his country and the richest mines in the world. On May 31st. this town was in British hands once again, beneath the protection of the British flag.

"**Pretoria. 17th. June, 1900**. *We left Kroonstad and crossed the Vaal River on the 25th. May and began the final part of our march to Pretoria. The night before we left, we heard the Highland pipers playing, "Soldiers of the Queen" and the men were singing, "God Save the Queen".*

We had several skirmishes with the enemy on our way up here, but nothing serious to speak of until we got near Pretoria on June 3rd. We were at it all day, but their determined riflemen, supported by cannon, held the approaches to the city; eventually, our guns and the big naval 50-pounder soon silenced their artillery.

As the Boers left, some of our Australian Mounted Infantry chased and captured one of their Maxims. On the morning of June 5th. we entered President Kruger's capital. We camped there for a few days during which time I fell ill and had to go into hospital, formerly the Transvaal Artillery Barracks. Here we have nice soft beds, such a treat after so many months of lying on the cold, hard ground. I am getting over my illness now and hope to rejoin my unit very soon.

Pretoria is a nice town, only just now its streets are rather dirty. It is situated between two high hills and there are five forts built around the town."

Edwin Moyle's illness was diagnosed as enteric fever. For three weeks he was very ill and could only swallow condensed milk, but he recovered: -

"*Enteric fever pulls one down so and I am now as thin as a whip-stick, but for all that I feel sure that I shall return home in as good health as when I left Melbourne.*"

Edwin wanted to return to the front but the medical officers sent him to Cape Town. Edwin's train went no further than Bleoemfontein because the Boers had blown up the lines: — *"I cannot speak too highly of the care and attention from the New Zealand nursing staff to us sick soldiers. I had the pleasure of a visit from the local Wesleyan minister and his wife who were very kind — it was my 21st. Birthday and they gave me a parcel from the church. Several of the Victorian Mounted Rifles are here with me, the rest of my regiment is on the road to Lydenburg. De Wet is still giving our generals a chase."*

PROFILE – TROOPER EDWIN MOYLE

Thus ended Trooper Moyle's war and he was invalided home, arriving back in Donald the same time as Arthur Hornsby. Our two returned soldiers had a "Royal Welcome" from the Donald folks. A further welcome was given to Edwin Moyle at the Jeffcott Mechanics Institute on the 21st. November when neighbours and friends presented him with a handsome, silver Rotherham watch, suitably inscribed; and then followed a concert given in his honour. Just before Christmas, Edwin was visiting Arthur Hornsby when he was suddenly taken ill and had to have constant medical attention for several days. During the next two years Edwin helped his father and also worked for Mr. Beckham. About 1903 his parents left Donald and retired to Learmonth. (Edwin lived with them because his four brothers had emigrated to New Zealand.) Some years later on the 7th. July, 1915, at the age of 36, Edwin Moyle volunteered to serve his country again. He was an experienced soldier, having fought for one year in the South African War and spent six years in the Victorian Rangers. He embarked from Australia on 23rd. November, 1915, and fought in France for three years with the 4th. Field Artillery Brigade as a gunner, and also as a driver of ammunition waggons carrying vital supplies across the horrific battlefields of the Somme. His sister, Eleanor Kathleen Moyle, served as a trained nurse in the field hospitals, and his brother, Gilbert Bennetto Moyle, fought as a rifleman with the New Zealand Forces. Later, Edwin married Elsie Maibacker, but there were no children. He and his wife worked on the Trewalla Station near Beaufort until he retired. Edwin Moyle died suddenly on his property at Ballarat, aged 64 years.

Boer war heroes return: Welcome home outside the post office, Donald, to Privates Hornsby and Moyle. November 19, 1900.

S.S. "Euryalus"

ON REACHING A RIVER

"During our day's trekking the sun had got extremely hot and there had been no opportunity of re-filling our water bottles, long since emptied. How glad everyone was to get to the water, and for those who got there first it was clear enough. Later on, when all the horses and mules, to say nothing of the men, had been into it, that was a different matter, and no doubt good advice as to boiling and filtering had been plentiful enough, yet time and opportunity of availing oneself of it was wanting, and for the most part everyone had to make the best of water of much the same colour and consistency as pea soup. It was easier to drink it in the dark."

ABOUT FLIES AND DUST

"Discomfort we suffered and diminishing rations, but a far worse trouble was flies. Flies in their myriads. Everywhere one looked the place seemed black with them, covering every morsel of food, crawling over one's face and hands, or drowning themselves in our rations of tea and making life to man and beast as unbearable as only such insects can. Sleep during the day was impossible and we had little at night. And we had daily duststorms which blew until it seemed as if there could be no more dust left in the whole of South Africa. Then the wind would change and blow the dust back to where it came from."

ABOUT TRENCH WARFARE

"We had a good look at his trenches after General Cronje surrendered at Paardeberg Drift (19th. February, 1900) upon the brown steep banks of the Modder River. The Boer laager itself was like a rabbit warren, pierced with holes and trenches in which the enemy had sought shelter from our artillery. The number of dead horses and cattle lying about it made it look more like a graveyard than a camp. The entire place was riddled with shrapnel and ignited lyddite. The trenches were really underground dwellings and perfectly secure unless a shell was dropped into the opening from above. Straight trajectory missiles were therefore unable to reach the defenders. **Such trenches had never been seen before in warfare. They were of great depth and formed veritable catacombs and tunnels.**"

ABOUT BRITISH KINDNESS

"The four thousand Boers who were now our prisoners seemed more pleased than distressed at Cronje's surrender. Reasons given for the surrender were that the Boers were running short of food and that Cronje, fearing that we would soon capture his position, and it being "Majuba Day", we might make free use of our bayonets. The Boers must have been glad and eager to leave their river trenches for in addition to the shells bursting upon them every minute, there was the most awful stench from dead animals and men who had been lying there for some days. When the Boer prisoners reached our camp, they were all given a good meal of cocoa, beef and biscuits. This was very generous because our own men were on half rations at the time, yet they shared their food with the starving prisoners."

ABOUT PRESIDENT KRUGER

"On the night of Tuesday, 19th. May, 1900, while the British army was surrounding Johannesburg, President Kruger with his secretary, Reitz, fled from Pretoria because the British were too close for his comfort. He took with him most of the coined and bullion GOLD from his country's treasury, leaving officials in all departments unpaid, or paid with useless "Treasury Notes". Both President and treasure fled away."

ABOUT SLAVERY

"Great Britain had abolished slavery in all its dominions in the 1830's which did not please the Boer inhabitants of South Africa because they treated the black natives as their inferiors and used them for cheap labour on their farms. Slavery was a big issue in South Africa because the Boers needed natives to wait on them, often beating them into submission; so the worst mistake Britain could possibly have made was leaving the black native to the mercy of the Boers after the war was over. We realise now that Great Britain should have made provision within the Treaty of Vereeniging to ensure that all natives would be given the franchise."

1st. January, 1900 — Distributing the Queen's chocolates

ABOUT RETURNED SOLDIERS

"It is a tragic sight, the return to Victoria in May, 1900, of the first batch of our citizen soldiers who are more or less maimed and broken in health in consequence of injuries sustained in the war on the South African veldt. To look at them is an object lesson to prove the indescribable misery produced by war. Six months ago these 49 brave fellows left our shores in buoyant spirits and with the blood running high in their veins as they contemplated an adventurous military career before them. Today, many of them find it impossible to walk without the aid of sticks or crutches, whilst others are lying helpless in pain on beds in Melbourne Hospital. Some will never be themselves again; shattered in limb and spirit they may never be able to resume the career at which they previously earned a living."

ABOUT LOCAL PATRIOTISM

"On the 31st. January, 1900, in St. George's Hall, there was a meeting attended by many Donald citizens for the purpose of collecting money in aid of the Patriotic War Fund. The money raised through various local concerts would assist soldiers who are fighting for the Empire in South Africa and also help wounded soldiers and their families. The great number of Donald citizens who turned out for the meeting was an indication of Donald's loyalty to the British Empire. The chairman, during his words of introduction, said, *'There are three fathers in our midst whose sons are at this very moment upon the battlefield. Privates FALLA, HORNSBY and MOYLE represent Donald and district in the Victorian Contingent.'* Three ringing cheers were given for these brave young soldiers by that vast assembly."

ABOUT THE BOERS' USE OF BRITISH UNIFORMS

"Often the enemy's success in small skirmishes was entirely due to their illegitimate use of British uniforms. One day a small band of English Yeomanry suffered heavy losses because the disguised Boers came so close. 'I thought,' said one officer, 'that I had made a mistake and been fighting our own men. They were dressed in our uniforms, and some of them wore the tiger-skin, the badge of Damont's Horse, round their hats'."

ABOUT ACTS OF TREACHERY

"One of the most brutal acts of treachery committed by the Boers was the cold-blooded murder of Captain R. MIERS, of the Somerset Light Infantry. It happened one day in September, 1901, when the English captain was riding his favourite grey mare, accompanied by his spaniel. He rode over to speak to two Boers who were waving the 'White Flag'. He dismounted and then walked with the two Boers behind a rock, out of view of his companions. Then witnesses heard a shot and saw his riderless mare galloping back towards the camp. When his friends eventually reached the spot, they saw his faithful dog sitting beside his body. He had been shot once in the chest and then stripped of everything except his shirt and a charm which he wore around his neck."

ABOUT THE LAST BOER WAR SOLDIERS

"The last surviving Boer War soldier in New Zealand died in 1980 on the 1st. April, aged 96. His name was BEN LEWIS SYMONDS and he sailed at 18 years of age from Christchurch with the Canterbury Regiment. The last surviving soldier of the Boer War in England was GEORGE IVES. He could pull his chin up to a parallel bar by the arms until he was well past 100, and complained that modern youngsters in their 80's and 90's were "apt to let themselves go'. The last surviving soldier of the Boer War in Australia is believed to be JAMES GORDON WILLIAMS, of Queensland, who died aged 107, in 1987. Even at that great age Williams could remember his slow voyage in April, 1902, to Capetown across the Indian Ocean with the terrible stench of the horses in the holds of the ship; and he can remember the dust and flies of South Africa as his contingent rounded up the last Boer rebels before the war finished on May 31st. 1902."

ABOUT THE VICTORIA CROSS

"During this Great Boer War, six Australians were awarded the Victoria Cross for outstanding valour. They were: NEVILLE HOWSE, JOHN BISDEE, GUY WYLLY, FREDERICK BELL, JAMES ROGERS and LESLIE MAYGAR. The first Victoria Cross in this war was awarded to Colonel CARLETON, of the Fusiliers, for his work at the Battle of Elandslaagte. The Victoria Cross carried with it a pension of £52 a year."

A RACE FOR LIFE

Although **Trooper PLANT** lived in the St. Arnaud district he had relatives in Donald to whom he wrote an interesting letter, recounting his amazing escape from death while serving in the South African Light Horse (better known as "Roberts's Horse"). This soldier learnt from bitter experience never to underestimate an enemy who is both clever and cunning.

Let the reader imagine a huge, flat plain that stretches for forty miles between the two towns of Thabanchu and Bloemfontein, with the Modder River flowing across this flat ground at a half-way point between the two places (see map). British troops were continually crossing this area so it would seem to be quite safe from the enemy, and on this particular day we are told there was not a single Boer **in sight**.

But the Boers are great masters in the art of ambuscade and there certainly **were** Boers upon that plain, about three hundred of them, concealed in a deep ditch at Korn Spruit, a tributary of the Modder River. They lined the sloping sides of this huge ditch with their rifles ready, waiting for the unsuspecting British column. What hope is there for a man travelling across a flat piece of land whilst his concealed enemy is pointing a Martini-Henry rifle at him? And one of the men on horseback riding straight into the Boers' cunning trap was **Trooper PLANT who lived to tell the tale.**

4th. April, 1900. *"We are now known as "Roberts's Horse" and a third of my regiment are Australians. Since I wrote to you last I have seen a lot of shot and shell as my regiment is with General French's "Flying Column", but the hottest time we have had was last Saturday (31st. March) when we were out on patrolling duties with General Broadwood's column.*

There were about 300 of us (Roberts's Horse) with some of the 10th. Hussars, Household Cavalry, Northumberland Fusiliers, New Zealanders and the Life Guards, with two batteries of Horse Artillery — about 1200 in all. We were camped beside the Modder River and during the night about 2,000 Boers came down upon us.

At daybreak on Saturday morning the Boers started to shell us and opened a brisk rifle-fire upon our camp.

We soon got the order to saddle up. Some of the lads opened up on them with rifles to protect our rear, and thus we made a retreat for Bloemfontein, retiring at a slow walking pace.

Some of our men were eating their breakfast (a piece of bread) on their horses, some were filling their pipes, and others were yarning. We were crossing a huge plain and there was not a Boer in sight. Our waggons were in front, followed by the Q and U batteries of the Horse Artillery which we (Roberts's Horse) were guarding. Behind us came the rest of the cavalry.

A heavy shell fire continued behind us, when all of

a sudden we saw the waggons forming into a laager (a circle) and we thought the Boers may have been spotted in front. Then suddenly some thousands of Boers became visible in a deep trench within a few yards of us with rifles in hand, and fifty yards behind their trench were two Maxim guns.

Our Colonel Dawson gave the order to "Files about! Gallop!" and we turned and galloped for dear life across the flat, every man for himself (sauve-qui-peut) but even while he was giving the order the Boers jumped out of the trench and fired upon us with one of the hottest deluges of fire that ever man saw — it was a regular hailstorm of bullets and the loss of our men and horses was heavy.

At first glance it looked as if nobody would live to tell the tale, but we rode on to a man determined not to surrender — death before disgrace.

I had gone about 300 yards when I saw a poor fellow running on foot and he looked fairly done in so I gave him my horse, but I soon got another one for there were plenty of riderless horses by now. I only got about fifty yards however when my horse was shot from under me.

I looked around for another, but could not see one to carry me out of that shower of bullets, so I lay down behind my dead horse, determined to shoot as many Boers as I could. I carried about 80 rounds and was determined not to waste a shot, but had only fired three bullets when I saw a splendid horse galloping my way.

I sprang up and ran for him as fast as I could, and by good luck caught him and jumped on and rode for my life. I reached my squadron and found they had formed up again. We sprang from our horses and flung ourselves

> down in a skirmish line and peppered away at the enemy, against whom we could only just hold our own — but we did it.
>
> We (Roberts's Horse) lost 100 men killed and wounded out of our 300, and I came out of it without a scratch. The name of the place where the fight took place is "Korn Spruit" or "Bushman's Kop".
>
> It is wonderful the kindness that is shown on the battlefield to the wounded. You will see men go back by the dozens into the thick of shot and shell to help their wounded comrades; and often get killed themselves in trying to save others."

The total losses in this disastrous, but not dishonourable, fight were severe. Apart from the number mentioned by Trooper Plant who referred only to the men in his "Roberts's Horse" General Broadwood's carelessness had resulted in a deplorable loss of lives. Altogether thirty officers and five hundred men were killed, wounded, or missing, and more than 300 taken as prisoners; and the Boers also captured a hundred waggons, a great quantity of stores, and seven 12-pounder guns. The Boers must have felt very pleased with themselves for giving the British "a surprise within a surprise" and for our poor British soldier it was certainly a war full of surprises, invariably unpleasant ones.

Why did General Broadwood not send his scouts ahead of the column to scour the ground? Had he forgotten that the Boers were masters in the art of the AMBUSCADE? Surely he knew that laying traps was a form of warfare in which his enemy excelled with great inventiveness, audacity and ingenuity?

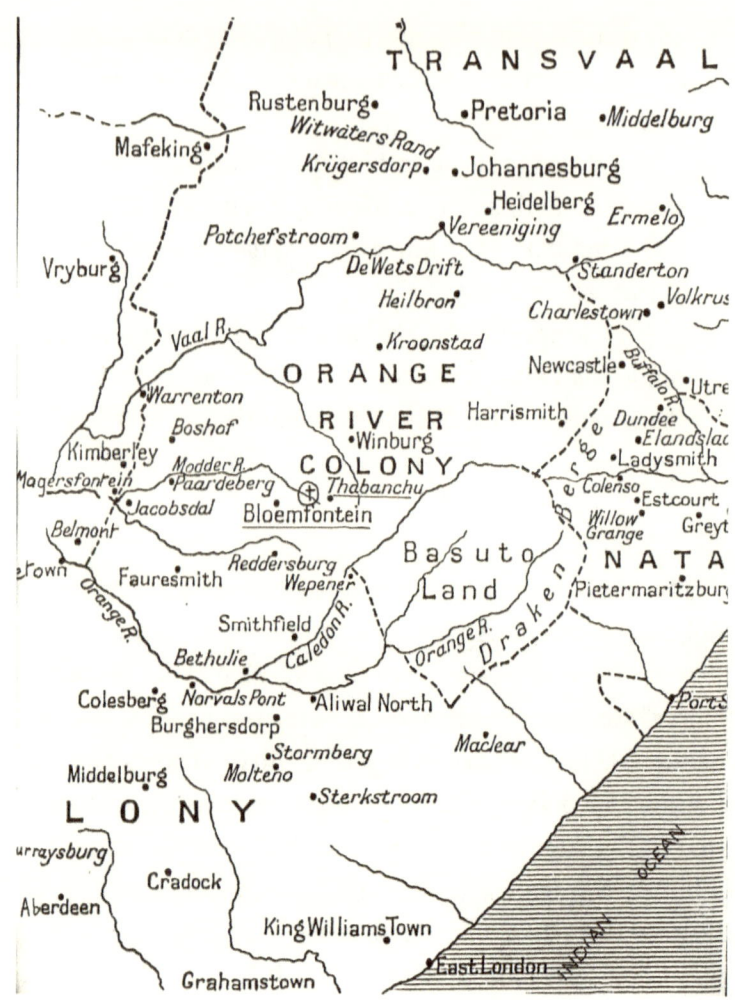

† = marks the spot where
Trooper PLANT escaped with his life

OUR BOYS IN SOUTH AFRICA

When its famous siege was over, the town of Mafeking was transformed into a huge army depot where the British columns came to replenish their supplies and it is here described by Corporal PERCIVAL MIDGLEY, a boundary rider from Minyip, serving in the Victorian Contingent of the Imperial Australian Regiment:

> "I am at present in Mafeking, having come in here for a rest. I have been here ten days and am returning to my column tomorrow. The reason I came here was to recruit my strength a little. I have had a double dose of malarial fever and now weigh scarcely eleven stone. I was out at Buffel's Hoek and the Captain ordered me to come into Mafeking. I did not want to leave my Company as we were having a little fighting every day. Malarial fever is not very pleasant to have — pains, aches and shakes all over, you can't eat and not much sleep and leaves you very weak. Our fellows are all longing for home and I think that is the case with every Australian. I will not be sorry to go back to my Company as Mafeking is a dirty, dusty hole; although it is not pleasant camping out on the veldt as it rains nearly every night. This town of

Mafeking is getting rebuilt after the effects of the siege. The inhabitants of Mafeking do not give such a glowing account of Baden-Powell as the world does. The niggers and Cape boys helped the besieged inhabitants a great deal and if it wasn't for them Mafeking must have fallen. I was rather surprised when I first sighted Mafeking as I expected it to be surrounded by kopjes, but instead it is surrounded by flat country. I am quite satisfied that the Boers lacked pluck or they would have taken it. There were eight thousand Boers against seven hundred defenders and that is big odds. I will send you some Mafeking flowers in my next letter as any souvenir from the renowned Mafeking is valuable."

RODERICK McSWAIN OF THE BUSHMEN'S CONTINGENT

At the amazing height of 6ft.-5ins. RODERICK McSWAIN was the tallest man from this district to go and fight in South Africa. Strong-looking and well-built, he developed his muscles during the course of his work as a blacksmith. Roderick was a popular man in the district and fondly known as "Rory". Before his departure to the war he was given a purse of sovereigns from his friends and Boolite neighbours, whilst the Donald community presented him with a fine pipe. Four Donald men, James Meyer, Roderick McSwain, William McAllistair and James Neyland, left Australia together as members of the Bushmen's Contingent, sailing from Melbourne on the "EURYALUS" on the 10th. March, 1900.

> **Krugersdorp. 23rd. March, 1901.** *"I have been pretty well all over South Africa and have been in several hot engagements. The worst was at Rhenoster Kop when we were 15 hours at it, only 500 yards from the enemy, and without shelter. The bullets fairly rained around us and I had nothing to eat or drink from 4 o'clock in the*

morning till dark; and the temperature during that day was over 90 degrees in the shade.

The Bushmen only had one killed and one wounded, but New South Wales lost four and New Zealand had twenty-seven wounded and killed. There were a hundred British casualties altogether, but the Boers' loss was heavier.

Although the enemy was hiding behind rocks, our shrapnel gave them "fits". We were only about 1000 men against 6000 Boers and they had a pom-pom, 7-pounders, big guns and Maxims directed straight at us. I can assure you it was an awful sight I shall never forget.

Our advamce was checked by their very severe fire so we dismounted and took such cover as we could find. The Boers opened fire on us again after we had unsaddled our horses, but they only shot a poor, unfortunate nigger who was sitting on a waggon. I had a very narrow escape when one pom-pom shell just missed my head. Every yard around us was commanded by their hidden riflemen and we could make no headway against such deadly fire. We just had to lie behind stones and anthills and endure their bullets until at last the sun sank. Then we had to saddle up again to go out and dig trenches and we continued digging all night. Next day after about an hour's fighting the Boers cleared out and we took their position.

A few weeks ago I sent in an application for the South African Constabulary. Shortly afterwards I was called in to the Captain and was appointed to the position of "shoeing smith" at a wage of 3 guineas per week. I have since been promoted to the post of Staff Farrier-Sergeant and am now receiving 4 guineas per week.

At present we are in camp at Krugersdorp, a very nice town about twice the size of Minyip. It is a very rich place as they dig up a tremendous lot of gold here. It is part of the famous Rand. Under its grasses lie such riches as even Solomon never dreamed of, yet we poor soldiers will not be one penny better off for seeing it.

This is the town where the Boers erected a huge monument to commemorate the disaster that befell the British at Majuba Hill on the 27th. of February, 1881. Each year, on Majuba Day, the Boers assemble here to indulge in making speeches and each person brings a small stone to place near the monument.

I suppose you wonder why the Boers have not been wiped out before now? Well, the fact is that this country in one mass of hills and hollows, and the enemy just keep moving about in small bodies from hill to hill; they will not stand still and fight "face-to-face", but fire a few shots behind your back and then clear off.

It is impossible to ride our horses up some of these hills as they are too steep, and consequently the Boers are away on some other hill before we can reach the top of the hill they were previously on. They behave like Bushrangers, not soldiers, roving from place to place.

Another thing is that it takes such a lot of British troops to guard hundreds of miles of important railway lines and therefore only some of us are free to chase the Boers. Naturally, the Boers know every inch of the country which gives them a great advantage in the chase. But Lord Kitchener will not take long to mop them up when he gets all his troops properly equipped.

The Australians have made a great name for

themselves in South Africa and have been highly-praised by all the British generals they have served under. I intend to stay here another two years as I can make money as long as I keep in good health. The climate of this part is excellent.

I have a memento of the war in the shape of a horse-shoe which I made out of the copper band from a "Long Tom" shell fired into Ladysmith during the famous siege. I have been offered £10 for it, but would not part with it for any amount."

Farrier-Sergeant Roderick McSwain had been on active service for twelve months when he wrote the above letter to his brother-in-law, Mr. James Penny, of Boolite. In it he described the deadly attack upon his mounted column by a superior force of Boers. Rory also points out to his reader that in spite of the thousands of soldiers whom the British had in South Africa, the vital lines of communication absorbed so many of them that mobile columns were often inferior in numbers to the enemy commandos who attacked them.

Roderick was patrolling with General Paget's force on the 29th. of November, 1900, at a place called Rhenoster Kop which was about twenty miles north of the Delagoa Railway line and fifty miles north-east of the Transvaal capital, Pretoria. General Paget only had with him about a thousand mounted troops made up of Australian, New Zealand and Tasmanian Bushmen, as well as the Yorkshire and Warwickshire Yeomanry with some naval guns.

The enemy was much stronger in number and well-entrenched in an admirable position on a ridge of Rhenoster Kopje from which they were able to keep up a deadly cross-fire upon our Bushmen, the New Zealanders and the Yorkshiremen. The fight ended with a typical withdrawal of the enemy during the darkness because the Boers always stuck to their motto -

"He who fights then runs away,
Will live to fight another day."

In another letter written from Waterval in September, 1900, Rory said he was now with General Plumer's force although he had been serving under Major-General Baden-Powell. Then he described their working conditions: -

"The heavy rains come in summer here and once when it was as dark as pitch I was guarding the waggons with another chap and had to drive two bullocks. The lightning and thunder were terrible and I could only see where I was going through the flashes. The rain poured down and I got wet through, but I stuck to my bullocks.

My skin is getting pretty well-tanned as I sleep with a thin water-proof sheet under me and a blanket over me out in the open, and sometimes the ground is so covered with small stones that I have to wriggle about for five minutes to make a level place and then I sleep as sound as a bell.

We are getting pretty fair tucker as we are good at "commandeering" fowls and pigs. We bayonet the latter, skin them, and make sausage meat of them, and don't we relish it.

I have still got my old draught nag and believe she

will pull me through unless she runs foul of a Boer bullet. My group mates' horses are all dead and there are not many of the horses still alive that we brought over from Australia with us."

A soldier's horse can never return,
On the veldt he must live and die.

PROFILE – RODERICK McSWAIN

Born on 3rd. October, 1865, and lived with his family at Boolite. His father was Donald McSwain, a blacksmith, who taught his son the same trade, so Rory lived at home and worked in the family business. In the official enlistment list it stated that Rory was "a single man, a Presbyterian and a blacksmith at Boolite". His height was quite remarkable (6ft.-4in.) and one can just imagine the comments from the enemy whenever they saw Rory charging at them with his bayonet. Rory had plenty of mates in the regiment and he is affectionately mentioned in some of their letters. After completing his twelve months of active service with the Victorian Bushmen's Contingent, Farrier-Sergeant R. McSwain was transferred to the South African Constabulary and soon gained promotion as a sergeant in that Police Force. In a letter to his sister, dated 24th. November, 1901, he wrote, *"The Captain wants me to sign on again, but I do not care about the way it is run, so I may or may not"*. However, he decided to stay with the South African Constabulary and was re-engaged for two years on the 14th. December,

1901; but first he took some leave and returned home to see his parents. Then in March, 1902, the local paper reported, *"Farrier-Sergeant R. McSwain of Boolite is about to leave again for South Africa. He will be entertained by his Carron friends at the Carron Hotel on Wednesday evening next."* Six months after the end of the war Mr. and Mrs. McSwain were officially advised that their son, Rory, aged 37, had died in South Africa from Enteric Fever on the 19th. November, 1902. He died in Rustenburg and was buried there in an unmarked grave; a tragic end to a fine soldier and an admirable example of Australian manhood whose career was so full of promise. His death certificate and effects were sent home to Boolite in March, 1903.

Lord Roberts enters Pretoria in triumph on 5th. June, 1900

Trooper Roderick Mcswain

THE SIEGE AT ELANDS RIVER —
AUGUST, 1900.

Conan Doyle wrote, *"This stand at Brakfontein on the Elands River appears to have been one of the very finest deeds of the war."* He was referring to the outstanding valour of 500 Australians who refused to surrender to an enemy that was five times their number. Our men were trapped and without water, their only gun jammed and their horses shot down, yet they were not disheartened and swore to die before the white flag should wave above them.

Trooper JAMES E. MEYER, a son of Donald's most respected founder, was amongst this brave contingent at the siege of Elands River and somehow he survived to tell his story: -

> *"Little did we expect to see in the grey light of early dawn 2,500 Boers hurling shells and bullets upon us. In the*

darkness of the previous night, while we slept, a large band of guerrillas under De la Rey gained possession of the surrounding hills which overlooked our garrison.

Rushing hurriedly from our beds we were greeted with a perfect hail of bullets and the contents of seven guns directed upon our little camp.

Shells were bursting into fragments around our heads and we seemed to be imprisoned in the middle of a hellish fire poured down upon us from the surrounding hills.

There we were, a mere handful of less than 400 strong and all Colonials — with the exception of a few Rhodesian Mounted Infantry.

Later, the Boer leader, Delarey, sent in word for us to surrender and complimented us on our brave stand against such odds. Delarey said in his message he would disarm all the troopers and give us enough rations to proceed back to Mafeking. Officers could keep their arms, but had to hand over all provisions and guns.

Delarey said if we did not do that he would blow us all to atoms with his 94-pounder gun. Our Colonel (Hore) sent back word that as the guns belonged to Her Majesty the Queen we would defend them."

Trooper GEORGE W. BOLDING, a cousin of Peter Falla, was also one of that brave number at Brakfontein on the Elands River, and his diary vividly records all the lurid details of the siege: -

"It was Saturday morning, about half-past six when we were preparing our morning scran and the Boers opened fire on us from all directions with shells, pom-poms, *explosive bullets etc. They had a good range, for from the very first their shells were uncomfortable for us.*

My mate, Frank Bird, had his leg shattered to pieces above the knee within a few minutes of the commencement. A friend, Corporal Norton, was wounded also by the same shell, poor fellow; he died before the day was out.

They kept a murderous fire on us all day long, bar at intervals when they would move their big guns to different positions. We had made bits of trenches around the camp a few nights before, but they were too shallow, not more than 15 inches deep (this was the consequence of poor bosses to look after us) and so when the Boers opened fire we had but little shelter.

To make matters worse there were over 100 bullocks impenned within ten yards of the trenches and a little further away were nearly 200 horses, all of which they started to mow down from the commencement.

Colonel Hore happened to be alongside our trench when the firing started so he was beside me all that day. He was in the siege of Mafeking with Baden-Powell and he said they had never had such a warm time of it during all those six months.

The Colonel called for volunteers to cut the bullocks loose so as to take their aim off us. Two of us jumped up and a nigger (the latter was killed while doing so). We cut them loose and drove them away, and the poor brutes were falling, killed and wounded, all around me. I can assure you I was glad to get back to our bit of cover.

Our losses on that first day were 6 killed and over 20 wounded. The total up to last night was 14 killed and over 30 wounded......

This is the eleventh day imprisoned in our trenches. We have made them nice and snug, four or five feet deep down and covered them over with bar loop holes and spying places to see how things are progressing outside. They are better trenches than the Boers can make.

We have not had any pom-poms or shells for four days now, but the Boers are still entrenched around us and keep having snipe shots. I don't think much of their shooting for we move about. Of course the majority of us take a little stroll when night comes and I go to the hospital to see how Frank and the rest of the wounded are progressing.

I have Percy Miles in the same trench with me and he is very much scared at times. Out of about 400 horses there is not more than 60 left alive. All the dead horses, bullocks and mules lie near our trenches, so you can imagine what it is like for the nasal organs."

Major DAVID JOHN HAM who came to the Donald district in 1875 to work on his father's selection at Mount Jeffcott, was in command of a Bushmen's Contingent at this time and wrote a very full account of their part in the Elands River siege: -

"All day long we lay in the scorching sun without a drop of water. That night our difficulties grew worse. We knew the enemy was guarding all parts of the river around us, yet we desperately needed water, so we decided to fight for it

A volunteer party took a few horses and crept down to the river's edge under cover of darkness. But enemy guards spotted them and there was a fight in which two men and several horses were killed. However, under cover of the firing, enough water was brought back to last us a day, each of us being rationed to a quart.

We had to get water in this manner every night after fighting all day. At daybreak on the second day, the enemy opened up with a hot fire from all their rifles and guns. Up to 10 o'clock we counted 600 shells. Then there was a lapse in the shelling, but their rifle fire was as hot as ever.

If our heads appeared above the ground for just one second, four or five bullets whistled in close proximity so that several of our men were killed in this manner.

On the third day their Boer General, Delarey, sent a messenger across to us carrying a white flag and a letter asking us to surrender. He congratulated us on our defence under such a devastating fire and promised he would merely disarm the men, but allow the officers to keep their weapons; of course he would take all our guns and foodstuffs and send us back to Zeerust (near Mafeking).

We all decided to refuse his offer and Colonel Hore sent back the message that we held the camp under higher instructions; and that we were Colonial troops of her Majesty Queen Victoria; and that we refused to surrender.

In response, General Delarey sent back his messenger threatening to blow us to pieces with his 94-pounder; **but at the same time he praised us by saying that the best fighting men he had ever met were the Colonials.**

For the next thirteen days we were in that place, cooped up like sheep in a pen waiting to be slaughtered. The dust was frightful and we had the option of washing our faces in the allotted quart of water, or drinking it. Needless to say, the water went inside.

> *Consequently we were unshaven, dirt-begrimed and powder — smoked beggars by the end of the second week, looking more like Boers than spick and span Colonials. Every man had a beard and was as black as ink from all the dirt and smoke.*
>
> *The weather was hot, the dead beasts were putrefying, and no relief came to save us. The sickening stench from dead horses, bullocks and mules, day after day, night after night, cannot be imagined; but it needs to be said so that you can have some idea of what we had to endure.*
>
> *Yet all this time our men were in good heart and refused to surrender until their last bullet had been fired and until each man's bayonet rested in the breast of his foe."*

Fortunately, on August 16th. this brave band was finally relieved by Lord Kitchener's convoy that happened to be passing nearby and heard the shots. When Kitchener sent General Broadwood's troopers to relieve that beleaguered garrison they were filled with wonder and admiration at the unbroken, defiant spirit of its few, emaciated survivors. Nearly all the horses and 75 of the men had been killed or wounded, yet that intrepid band never gave up hope, not even after a failed rescue attempt by General Carrington and his troopers on August 5th.

Trooper GEORGE W. BOLDING wrote about the end of the siege in his diary, under the date 18th. August, 1900: -

> *"We were relieved by Kitchener, General Ridley, General Broadwood and two other generals (whom I forget their names at present) with Mounted Cavalry, Lancers, Infantry etc. their numbers amounted to 10 or 15 thousand men, such a sight and a very welcome sight it is.*

We had officers and men of all descriptions strolling around our camp and talking about the fight we had and comparing it to what they have had. One chap said he had been in 46 engagements and he had never seen such a mess made by shells and bullets before. They are all surprised to see us so happy. One officer reckons we should get two bars instead of one.

We heard there were close on 3,000 Boers with eight guns, firing over 2,000 shells on us that first day.

I must say the Boers treated our sick lads very well. Frank Bird is progressing very well. Indeed I miss him very much; he used to do all my cooking and such like. We had a burial service this morning over the poor chaps who went under during the seige.

There is still fighting on both sides of us. I am as happy as Larry."

Trooper GEORGE W. BOLDING died some weeks later from the dreaded Enteric Fever, due to drinking polluted water during the siege. Trooper JAMES E. MEYER and Major DAVID J. HAM also caught Enteric Fever, but they were more fortunate and managed to recover.

"If you want us, come and get us!" yelled a defiant defender.

THE SIEGE AT ELANDS RIVER — August, 1900.

THE SIEGE AT ELANDS RIVER

Full of admiration after hearing about the brave deeds of our colonial men in the Great Boer War, ARTHUR CONAN DOYLE wrote: -

> "The Australians have been so split up during the campaign, that though their valour and efficiency were universally recognized, they had no single exploit which they could call their own. But now they can point to **Eland's River** as proudly as the Canadians can to Paardeberg.
>
> They were 500 in number, Victorians, New South

Welshmen and Queenslanders, with a corps of Rhodesians, led by Colonel Hore. Two thousand five hundred Boers surrounded them, and most favourable terms of surrender were offered, but bravely scorned. Six guns were trained upon them and during eleven days 1,800 shells fell within their lines ..... but they were sworn to die before the white flag should wave above them.

When the ballad-makers of Australia seek for a subject, let them turn to Eland's River for there was no finer resistance in the entire war."

PATRIOTIC SONGS OF THE SOUTH AFRICAN WAR

Richard Merrett, BoerWar Soldier, 1900

CAPT. D. J. Ham (B DIVISON).

TROOPER WALTER COOMBS

Some British soldiers could ride and some could shoot, but very few could do both. No wonder when our volunteers arrived in South Africa they were immediately accepted into the mounted infantry units. Australians were used to working outdoors in the heat and dust as boundary riders, drovers, farmers or bushmen — and all had learnt to shoot straight from the time they were big enough to hold a gun. Also, the terrain of South Africa greatly resembled the Australian outback, so it was in the Australian soldier that the Boer at last found his match for the Australian was a fighting man, a lover of independence, a free thinker "who gave his officers a headache, but gave the enemy an even bigger one". Eventually, our bushmen were admired and hailed as the best mounted cavalry in the world who could handle any horse.

WALTER COOMBS, from Corack East in the Donald district, was a trooper in the West Australian Mounted Infantry, and he describes some of his experiences in a letter dated 9th. July, 1900: -

> "We are in Bethlehem (a strange name to be connected with war) and it is in the Orange Free State. The village was captured by us on July 7, but we are in the midst

of a bad war zone and we have been under heavy fire continually for the last fortnight; having heavy fighting also means heavy losses. My mate has been severely wounded by an explosive bullet through his breast

I have escaped so far, but had one or two narrow shaves, one being a shell that burst about 40 yards from me and a piece struck my saddle but doing little damage beyond a severe shock; as for the bullets, well they just buzz over us like a swarm of bees.

We have had to fight our way every day bringing relief to a town called Lindley which would have surrendered to the Boers but for our arrival next day. It is sad to see the desperation of the English inhabitants who are driven to eating maize-meal (horse food) and have nothing else.

In fact it's not all fun to be a soldier. We only get about three biscuits a day as hard as iron, and army rations consist of meat and vegetables, all tinned and cooked (a 2lb. tin between two men) but we always catch sheep, fowls or geese to supplement our rations.

Hoping this finds you all well in Corack; remember me to all friends and tell them we have not caught Kruger yet, but he can't hold out much longer as he is nearly cooped up now. I may yet see you all again, but I must tell you that a man's life is not worth much here for we are in the midst of death nearly every day.

The other day when we beat back the enemy we finished up with a bayonet charge which drove the Boers out; they run when they see our bayonets. Of course, we had the artillery with us which did splendid work shelling the Boers from hill to hill.

We had a shock one day as we were shelling the Boers'

position. Unseen by us and hidden by the tall plants, they crept up through some maize fields and shot every gunner and officer on the guns; in fact they had their hands on our guns when we charged them and drove them back. We killed 70 Boers and only lost 7 men, but we rescued the guns, so you see it was a close call.

What puzzles me is how we shall come out of it all? Only those who have been under fire on a modern battlefield of today can form any idea of the effect of shells and bullets and noise; but one gets to hear music in the squeal of a shell and the hiss of a bullet, at least that's how I feel."

Trooper Walter Coombs and his squadron were few in number, yet they courageously wrenched the Royal Field Artillery's guns back from the very hands of the enemy. This small band of Australian Mounted Rifles had been escorting a large British column, led by General Paget, during a difficult and dangerous march through the Orange Free State. They had been harassed by enemy snipers every inch of the way and their advance upon Lindley and Bethlehem was particularly dangerous because De Wet and his men were lurking in the surrounding hills.

It therefore was inevitable that this British column would eventually encounter a large Boer commando — and it happened on 3rd. July, near Leeuw Kop. How modestly Trooper Coombs describes the part he played in the ensuing battle! It was thanks to his small band of Australians who rushed to the rescue of the 38th. Royal Field Artillery that the outcome of the battle was reversed. In the face of sharp rifle-fire Trooper Walter Coombs and his mates attacked and recovered the ground lost to the Boers and then with their bayonets drove the Boers away from the British guns.

> "GREAT DEEDS ARE DEATHLESS THINGS;
> THE DOER DIES, BUT NOT THE DEED."

PROFILE TROOPER WALTER COOMBS

Trooper Coombs was born at Corack and spent his youth in the bush where he learnt to ride rough and shoot straight. In 1900 he volunteered to join the West Australian Mounted Infantry and eventually embarked for the South African War. On his safe return in September, 1901, he was extended a warm welcome by the residents of Corack who celebrated the happy occasion with a concert, supper and dance in his honour. He returned to work as a farmer on his father's property until fourteen years later when he volunteered once again to fight for the cause of Freedom. He gave his life in the First World War when he was shot down by the Turks on the 25th. of April, 1915, during the Landing at Gallipoli.

Walter Coombs
He Fought In The Boer War And Was Killed In The First World War

JAMES E. MEYER
— A SOLDIER OF THE QUEEN

At the beginning of February, 1900, James E. Meyer passed the riding, shooting and medical examinations for enrolment in the Bushmen's Corps. Then he went to the Military Training Camp at Langwarrin to begin his serious training in readiness to serve in South Africa.

He returned to Donald for a final farewell and on the 22nd. February, residents and friends gave him a great send-off at the Royal George Hotel.

The chairman of the proceedings upon this auspicious occasion said, as he presented James Meyer with a Paterson's Patent Pipe and cleaner attached: -

> "James is going overseas to fight for the cause of Freedom and for the Honour of the British Empire in South Africa. We hope this pipe will be of great comfort to you throughout the war, and that you will have the health to smoke it and still be alive and well to bring it back to Donald."

Two other men of the Donald District were leaving with James Meyer in the Bushmen's Corps for the Great Boer

War, namely Roderick McSwain (Boolite) and J. Neyland (of Corack) so James was asked to take two more pipes, similar to his own, and give them to Mr. McSwain and Mr. Neyland with the compliments of all Donald residents. When thanking the Donald people for their generous present and good wishes James replied: -

"I am a heavy smoker so will really appreciate the smoking of such a fine pipe; and I am sure that my fellow comrades, McSwain and Neyland, will also be thankful to you for this kindness shown to them. Our dear old Colonel Otter warned us that it is no child's play going to South Africa and that we would have to work hard and perhaps be in the saddle for two days continuously; and when ordered out to draw the enemy's fire we will be told not to return the fire. Should we disobey orders, we will be shot! I am not much of a speaker, but Donald people may rest assured that I shall do my duty for Queen and Country."

The 3rd. Contingent was known as the 'BUSHMEN'S CONTINGENT' because its members were farmers and bushmen "who were hardy riders, straight shots, accustomed to finding their way about in difficult country, and likely to cut an expert figure in the vicissitudes of this South African campaign".

While serving with his unit of Mounted Riflemen, Trooper JAMES E. MEYER kept a little diary in which he wrote some facts about chasing the Boers: -

We are riding night and day, up at two in the morning and don't go to bed until 10 o'clock at night. We do not get half enough sleep and the men are all fagged out. It is a brute chasing Boers over the hills and we travel for miles through rocky country. The trouble is that the Boers move so quickly; there may be none in the hills one day and then the next day they will suddenly appear on the horizon. I tell you that a man on a horse riding across the plateau has no chance against a Boer who is lying under a rock on the top of a steep, stony kopje. After all, a single Boer can shoot you just as dead as a hundred Boers could in a battle.

We lose two or three horses each day through weakness and lack of forage. Sometimes the tracks across the veldt are so dusty that it is impossible to see a comrade who is riding beside you! There are terrible swarms of locusts for miles around and they cover the ground like a thick carpet. As we ride through places en route and stop at the stores, we find shopkeepers have sold out of everything; they charge us double the price anyway.

One day we passed a monument erected in memory of 44 settlers murdered by the natives in 1876, a very nice stone with all their names inscribed on it.

The other day I shot a python snake, 12 feet long and as thick as my leg, capable of swallowing a small deer. We had rum served out to us in honour of the Queen's Birthday on the 24th. of May.

James wrote the following letter to his widowed mother telling her about his adventures: -

2/6/1900. "When we reached a place called Bamboo Creek, twenty of us were ordered back to camp in Biera to attend the horses which were dying at the rate of five each day, of a disease called "Blue Tongue" which takes

Mr. John August Meyer And Family, At Inglenook, Donald In 1890. Lena, Gus, Mr. John Meyer James, Mrs. Charlotte Meyer Oscar

them off within half-an-hour. It is generally the healthiest horses that suddenly die. There is a fortune for any man who can find a cure for it, yet men all over the world have tried to cure the disease but none has succeeded yet.

At Bamboo Creek the disease is worse than in Biera and if it continues we will not have a single horse left to ride. There are 7,000 more Hungarian horses to be landed here for the use of our troops. They are a small breed of horse which I don't think we could breed in Australia.

The trip up to Bamboo Creek was well worth seeing; through jungles and across veldts covered with beautiful flowers and plenty of game of all kinds — monkeys, deer, zebras, wild dogs, hyenas and lions which we can hear roaring at night.

It took us eleven hours to travel 68 miles by train and we stopped six times to put the overturned trucks back on the lines; so that will give you some idea of what the railway is like in South Africa.

Some of the boys are down with fever. I am surprised that more are not ill for this is one of the worst fever places in Africa.

To make matters worse, we have just had three days of heavy rain. White people cannot stay here longer than three months without catching a fever. Niggers are dying every day.

I am here with the New Zealanders, Queenslanders, Tasmanians, Canadians, West Australians,

New South Welshmen and South Australians, but we have not the slightest idea when we will go to the front.

We have to guard the horses day and night for this place is full of thieves and spies. Our food rations are poor, just one pound of tinned meat and six hard biscuits a day and also one pound of jam for twenty men, three times a week.

I had one day off to go into Biera to see the town and a miserable place it is, so we call it "The Gates of Hell". You cannot walk along its streets without sinking into sand about two feet deep, so the white inhabitants travel in a little carriage which is pushed by the niggers along narrow tram-lines. The poor niggers are nothing but slaves to these Dutch (Boers).

My Bushmen's Contingent played the Biera Cricket Club last Sunday and beat them by 28 runs and so the Biera Club entertained our team to a banquet in the evening. The white inhabitants here only dine twice a day with one meal at 11 a.m. and the other at 8 p.m.

We are all sleeping in cattle trucks and water is very scarce. Sometimes we don't get a wash for three days. All water has to be boiled before we can drink it.

I have got my own mare looking well, but I am afraid it is labour in vain for I expect she will die of this awful disease at any moment. It is a pity to see such noble animals dying like this, although it is a good job they don't suffer for long.

A few of us crept out of camp the other night and saw a Koma dance by the niggers which takes place once a month under the full moon, and what a great sight it was!

The New Zealanders have fine horses and some of them are the best ever landed, but horses here are fed on nothing but bare oats. Fever has broken out amongst the men again and two more of our lads have died."

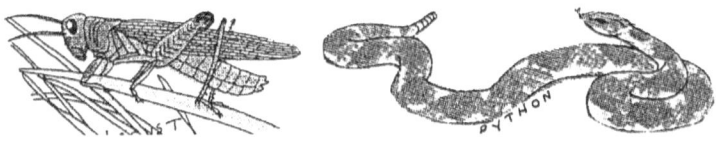

A tragic accident happened to one of the Bushmen as they journeyed by train from Umtali to Marandellas in May, 1900. It seems their troop-train had broken down, so Sergeant Brent went up to the driver and kindly offered to help him. Suddenly, another train going in the same direction rammed into the back of the stationary troop-train, knocking Sergeant Brent onto the line and mangling him up. The poor chap was buried at Umtali with military honours.

During the next two months, Trooper James Meyer rode with the Bushmen as they fought their way through large numbers of the enemy as far as Elands River. The story of how they were attacked and surrounded by a superior number of Boers has already been told; however, the following statement in a Mafeking newspaper dated 1st. September, 1900, makes interesting reading: -

A PLUCKY BAND OF AUSTRALIAN VOLUNTEERS

The brave little garrison who stuck to their post at Elands River so manfully, and thereby gave the world another example of how British pluck and determination can come out on top in the face of overwhelming odds, have arrived safely in town. They came in with Lord Methuen's column and in consequence of their horses having fallen victim to the Boer shells, these heroes had to march all the way on foot. Their garrison was being besieged for thirteen days. They had camped beside a kopje commanding the main road, peacefully passing away the time while awaiting orders to proceed to Rustenburg. No attack by the enemy was anticipated; still, to be in readiness for any emergency, the men had built small semi-circular sconces (protective earthworks) of loose stones about 2 feet 6 inches in height and 15 inches thick. On the night of August 4th. they laid down to sleep in the belief that patrols would go out on the morrow as usual and that there would be no alteration in their duties from those of the preceding few days. But with daylight next morning came proof that their calculations had been terribly wrong. During that night the Boers had surrounded them and positioned their guns on the high kopjes to the north, east, west and south, around them. In addition to this, the Boers had made preparations to play their favourite game of "sniping" by placing a large number of sharp-shooters in favourable spots to pick off our men. Early in the morning the attack began. How "mild" may be judged from the fact that 80 big shells were poured into the camp in two hours, and over 2,500 shells were fired upon them during the first two days of the siege. The Boers took a special delight in sniping at the men while they were watering their horses . . . . etc.

Many of that "plucky band" later succumbed to Typhoid or Enteric Fever because they drank polluted river-water during the siege. Trooper Meyer had Enteric Fever and was very ill in hospital for several weeks.

He was invalided home to Australia and reached Melbourne, via Hobart, on the 14th. of December, 1900, after what he describes as "a cold and uncomfortable voyage". He was met at the docks by his mother, Mrs. Charlotte Meyer, and his youngest brother, Oscar Meyer.

On December 17th. a concert was held in St. George's Hall for the purpose of welcoming back Trooper James Meyer who briefly outlined his career in South Africa. Then on the following Wednesday evening Mrs. Charlotte Meyer entertained a large gathering of over 150 friends in honour of her son's safe return from the war.

However, less than one month after Mrs. Meyer's grand reception for her son the local newspaper reported that volunteers from Donald for the Federal Contingent were: — R. T. Muir, C. Routley, J. Hamilton and James E. Meyer.

James Meyer was invalided home from South Africa only a short time ago, but he has gone to Melbourne to submit to a medical examination and if it is satisfactory he will shortly return to the front. ("Donald Times" — 15/1/1901)

Lieutenant James Powell and Horses at Port Melbourne

"The Obstinate Heart shall be Laden with Sorrows."

LORD ROBERTS: "Stand aside, Madam, I have listened to you long enough. This miserable business must now be ended, and quickly!"

To counter guerrilla attacks upon his soldiers, Lord Kitchener began a two-pronged policy which was to cause deep-seated anger and bitterness between the British and Boer nations. Kitchener's plan was intended to deprive the roving Boer commandos of food and shelter to be found at certain disloyal Afrikaner homesteads, while at the same time rounding up Boer women and children. Where it was suspected the rebel commandos had sought shelter and food, farmhouses were razed to the ground and all the farm-buildings destroyed along with animals, crops and orchards. Refugee camps were generously set up by the British Government to provide food and shelter for any homeless civilians, especially women and children. Concrete "blockhouses" with manned guns were erected across the country close to the railway lines, and these were linked by miles and miles of barbed-wire fencing into which the rebels could be driven and caught.

WILLIAM RUSSELL WALDER
"WEDDING DAY"
Photo: Ian Walder collection

GUERRILLA WARFARE

WILLIAM RUSSELL WALDER was born at Laen in 1885, just before his parents moved to a farm near Watchem. He was the second son of pioneer settlers, James and Sarah Walder, and grew up in a family of ten boys and five girls. Being able to ride and shoot were necessary skills for survival in those pioneering days and so William was taught these activities from a very early age.

William was only fifteen years old when the South African War began in 1899. He was just seventeen when he decided to go and fight for his Queen and Empire, but to avoid rejection in Victoria because of his youth he paid for a ticket on a boat that left Port Melbourne on the 16th. of March, 1901. He arrived in Cape Town on Tuesday, 16th. April, and wrote in his dairy: -

> There is plenty of dust about this place and the streets are very poor, nothing to what I expected to see. I went to the drill hall to find the recruiting office and enlisted in the **2nd. Scottish Horse**. I signed on at 11 a.m. and by four o'clock that same afternoon I was in Maitland Camp, in uniform, and with all my kit.

Scarcely one week after joining the 2nd. Scottish Horse, William was sent to the front. It was a long train journey, traversing the length of South Africa through Cape Colony, Orange River Colony and eventually into the Transvaal. The train stopped for a few hours at Norval's Pont (where he enjoyed a swim in the Orange River) and then passed through Springfontein, Bloemfontein, Geneva and Klip River, eventually reaching Elandsfontein, Trooper Walder's destination. Here he saw 10,000 troops, including 250 men of the Scottish Horse.

At Elandsfontein Camp there were plenty of parades and drills to keep the troopers busy. They were shown how to take good care of their horses which were few in number, many having been killed in battle. There was a splendid reservoir about a mile from their camp where the men could go and bathe; and in the nearby town were three Soldiers' Homes to provide them with such pleasurable pursuits as reading, games and afternoon teas.

British military historian, Arthur Conan Doyle, speaks highly of the force in which Trooper WILLIAM WALDER served during 1901: -

> *"Of all the sixty odd British columns which were traversing the Boer states at this time there was not one which had a better record than that commanded by Colonel Benson. During seven months of continuous service during 1901, Benson's small force consisted of the Argyll and Sutherland Highlanders, the **2nd. Scottish Horse**, the 18th and 19th Mounted Infantry, and two guns. Leaving the infantry behind to guard the camp, Benson always operated with his mounted troops alone*

until no Boer laager within fifty miles was safe from their nocturnal visits."

While stationed in the Elandsfontein Camp, Trooper Walder had plenty of practice doing escort duties — sleeping out on the open plain with no protection. In the evenings the only entertainment for troops was an open-air concert with about 3,000 men present, chaired by the camp commandant. (There was plenty of acting and singing talent amongst this large number of soldiers.)

Guarding the nearby gold-mines was also a necessary duty for some of the mounted troops; and whilst breaking-in horses was an easy task for Trooper Walder, it was very difficult for new chums just arrived from Scotland who were poor riders when sitting upon lively animals: -

It is good fun watching them trying to ride. All the bad riders have to do extra drill, bare-backed, every day.

About a month later, on the 6th. of June, Trooper Walder's squadron was sent to Machadodorp to join Colonel Benson's column which was about three thousand strong; and when Major Murray inspected this new squadron he congratulated them on being such a fine body of soldiers. He also gave them a lecture about the necessity of looking after their horses during long treks.

Guerrilla warfare was raging throughout the Transvaal. From Lichtenburg to Komati there was sporadic, but deadly fighting. Strong and mobile Boer commandos with guns moved about in the area and a British soldier's greatest fear was a bullet in his back from hidden, enemy snipers. British authorities knew that as long as Boer leaders like Botha, Delarey, Steyn and De Wet remained uncaptured the flame of resistance would continue to burn.

Speed and surprise were essential elements used in the capture of Boer commandos and Colonel Benson was master of such tactics. His "Flying Column" was kept very busy doing night-marches and dawn-attacks.

Colonel Benson's method of surprising the Boers was simply for him and his men to gallop headlong into the Boer laager (camp) and to go on chasing as far as their horses would go. The furious and reckless speed of his mounted troops in such an attack can be judged by the fact that more of Benson's cavalry were hurt from falls than from bullets. Trooper William Walder's diary tells us: -

> **JUNE, 1901**. *Machadodorp is a pretty little town and a very large military centre. There are hundreds of tons of provisions stacked up here, both for men and horses. The Victorian Squadron arrived last week and I saw my old schoolmate,* **Jimmy Muir**. *The column is packing up ready to move off early tomorrow so we pulled down all the tents and will have to sleep out tonight... It was a spendid sight to see the column moving off. This country is terribly rough, nothing but mountain-climbing all the way and we often pass the remains of burnt farmhouses... I like trekking alright, but we are camped amongst stones tonight and it is a job to find a smooth place to lie down, but I suppose we will get used to that... We stopped at Lydenburg, about the prettiest little town I have seen over here... From Pilgrim's Rest two of our squadrons went left while we went right to try and surround a*

Boer laager. We were sent out among the hills to try and cut off the Boers' retreat, but we rode about three miles and then dismounted because most of our horses were lame; and we walked about ten miles over hills and across valleys all day. I saw the most beautiful waterfalls, and we went through some small forests so dense that we had to cut our way through... Today we left camp and marched out through the hills, burning farms and bringing in Boer families with all their farm stock.

June, 1901. *We had two men wounded on outpost today... We marched out of Lydenburg this morning and came about seven miles over very rough country. We passed some very fine orange groves, so everybody filled their nose-bags and haversacks with oranges and we then camped in sight of the main column. There is a beautiful little river running past the camp so we all indulged in a good bath... While we were away from the column there were two men killed, an Argyle and Sutherland Highlander and a Mounted Infantryman... Today, we had a sharp engagement with a commando of Boers under De Beer. They escaped, but left behind the whole of their convoy, stock and provisions... The column is staying at Clootsfarm. While we were climbing amongst the hills we found a few Boer waggons hidden in a kloof (ravine) and we collected a few Boer families, plenty of their poultry, and a hundred head of cattle. We had a bit of a scrap with a few Boers who were sniping at the rear of the column, but no casualties... We are resting at Bootsfontein today with plenty of wood and water, so we are cooking poultry and pork for a regular* *feast tonight... From Clootsfarm we were sent out to search a valley where we think the Boers have hidden their stores. There are hundreds of baboons and wild deer in the kloofs. We found about a hundred cattle, six waggons and some Boer families hiding in a kloof. In a cave we found provisions and clothes... The column is on its way back to Machadodorp for a rest and to wait for a lot of remounts (new horses) because about a hundred of our men now are walking, their horses knocked up or dying. All the horses are very weak.*

The British still have not learnt the obvious lesson — that it is better to give 1,000 men two horses each and so let them catch the enemy, rather than give 2,000 men one horse each with which they will never **cut off the legs of the Boers!**

July, 1901. *We are resting at Macadodorp today, busy washing our clothes in the river running past the camp. There is little feed for the horses on account of the large numbers of Boer cattle being sent here from the five columns operating in the area. We are expecting to ride tonight against a Boer laager reported to be about twelve miles away... We rode out of camp at two o'clock this morning and marched in silence till daylight, then took up positions on kopjes, expecting a general engagement. One of our squadrons attacked, but the Boers replied strongly with their guns so we were sent to reinforce our men. The fight lasted two hours and "H" Squadron lost heavily with three killed, five wounded and several horses shot down, while the Boers had one killed and several wounded. Our men that were killed were buried on the* *field. Our pom-pom (a Maxim automatic quick-firing gun) arrived too late, just as the Boers were riding out of range. This is the place where Lieutenant English won the Victoria Cross... Our column halted at Elands Kloof while Mounted troops were sent out to collect families and burn their farms. We found rifles and ammunition buried at one farmhouse... We got plenty of poultry and goats and burnt quite a few farms today. The weather is very cold but no rain. We are still chasing Commandant Viljoen who is using two pom-poms captured from the British and his commandos are busy in the Lydenburg and Dulstroom districts. Around here the countryside is very rough, but a great place for freshwater springs... We left the old camp in Dulstroom at seven o'clock and at ten o'clock Viljoen opened fire upon our left flank with his two pom-poms, but did no damage. All the artillery was brought into action and he had to retire. We chased him about eight miles until night came on and then we retired...*

In his diary, Trooper William Walder describes how he chased Boer commandos across the Transvaal during the winter of 1901. Mobile Boer commandos were successfully harassing British forces within South Africa, so Lord Kitchener was forced to become increasingly harsh in order to stop this seemingly endless war. Consequently, Boer farmhouses were razed to the ground; homeless families placed in concentration camps; and Boer prisoners deported to the islands of St. Helena and Ceylon.

During his work as a Mounted Infantryman, Trooper Walder took part in several night-marches followed by dawn-attacks upon Boer laagers. He burnt down farmhouses (where Boer commandos might find food and shelter with sympathetic burghers) and he commandeered (and enjoyed eating) the farmer's poultry and pigs.

> **July, 1901.** *We were sent out in a patrol of two hundred with a pom-pom to clear a few farmhouses. We found two wounded Boers in a church and they said Viljoen was very hard on them... Our Mounted troops and a battery of artillery left camp at two o'clock this morning and the convoy followed on at daybreak. We found Viljoen and chased him for three hours, killing two of his men, wounding three, and taking three prisoners. We also captured a large portion of his convoy. We had three men seriously wounded... Another day, chasing Viljoen and severely shelling his rearguard, but he kept going till nightfall and we had to retire. He is very hard pressed and we expect to catch up to his convoy in another couple of days... About midnight, after ten miles of hard riding, we found Viljoen and his convoy in a kloof (ravine). We surrounded the kloof and wounded seven of his men, but Vijoen and the rest managed to escape. We captured 17 waggons, 40 Cape carts loaded with stores, 30 horses and saddles, 90 women and children, and a few rifles that the Boers dropped in their haste to get away.*

These captured women and children were taken to the concentration camps which had been established by Lord Kitchener with the very best of intentions. Women and children were collected from the ruined farmhouses in this devastated countryside and taken away to concentration camps where they would be provided with food and shelter. This was a generous and humane movement on Kitchener's part, although by the end of the war 4,000 Boer women and 16,000 Boer children had died in his camps from infectious diseases such as measles, pneumonia and enteritis.

Her enemies were quick to blame Britain for all these deaths; however it must be said that the sanitary arrangements of the Boer people were of the most primitive description because of their total disregard for elementary cleanliness, and their slovenly, insanitary habits within a crowded refugee camp would have been important factors in the spread of dangerous, contagious diseases.

Trooper William Walder spent most of the African winter moving incessantly across Northern Transvaal, hard on the heels of the notorious Ben Viljoen. He assisted in the clearance of Boer farmhouses, waggons, provisions and farmstock, but no one could catch the slippery Viljoen and his band of rebels. They were as cunning as sharks, remaining submerged until it was time to strike some unsuspecting British column.

Boer guerrillas understood the principle "that invisibility is the best means of survival"; and although their "hit-hard-and-run-away" tactics might appear to be cowardly actions, in practice this meant that they lived to fight another day and therefore lost fewer men.

July, 1901. *The wind rose this morning and it was very dusty all day. My mate was sent to hospital with dust on his lungs. We left camp this morning and marched out through the hills, burning farms and bringing in families and stock. We commandeered plenty of poultry, pigs and* *goats at these farmhouses... When the column halted again, Mounted troops were sent out to collect families and burn the farms... Some Boers attacked our rearguard, but did not come too close. We made a long and forced march after leaving Blood River Valley, and the road was very dusty. The Boers were about two days' march ahead of us so we left Wagendrift camp early; but then suddenly the Boers attacked us with rifle and Pom-pom fire.*

In less than fifteen minutes our Captain, Lieutenant and a Corporal were killed and some of our horses shot from under us. Then our artillery came into the action and shelled the enemy for about two hours before we could advance another inch, and then the Boers were forced to retire leaving behind a number of their dead and wounded. We crossed Blood River — a large river with a very strong current — and after riding eight miles we camped for the night, but did not see Viljoen again.

Trooper Walder was just one of many British soldiers who were chasing Boer guerrillas over the Transvaal and the Orange River Colony during the winter campaign of 1901. The total number of columns engaged in this work amounted to at least sixty and in each column there could be any number of men from two hundred to two thousand — but these columns seldom hunted alone.

July, 1901. *We left Waterval and marched seven miles today and camped alongside the Fifth Victorian Contingent. They were very pleased to see us as they have had a hard time. Our poor horses were very* *thirsty, not having had any drink that morning. We held sports between the two columns in the afternoon... The weather is warm and the road dusty. We passed a very nice Kaffir town today and it had about a hundred kraals and a tremendous number of blacks... We arrived at Middelburg today after a long trek over very rough country. The horses are very much done up and want a rest, and so do we. There is another column camped not far from us so the two columns had a combined sports consisting of horse and foot racing of all kinds. Colonel Benson gave a prize of five pounds for the quickest turn-out; it was won by the Eighteenth Mounted Infantry. We are going to do a lot of Flying Column work and night marches soon. There is a company of West Australians with us now.*

In the meantime, Boer leaders carried on a vicious campaign using nasty guerrilla methods; and like heavy bull-dogs chasing swift greyhounds, our weary soldiers trailed behind the armies of Botha, De Wet, Delarey, Smuts and Viljoen. These rebels knew the lie of the land. They rode sturdy Cape ponies and collected necessary food and clothes from farmhouses along the way, or "borrowed" uniforms from captured British soldiers.

August, 1901. *Mounted troops left camp at seven o'clock last evening with two big guns and a Pom-pom and we rode about twenty miles. At daybreak we surrounded a small laager and captured seventeen prisoners including a Field Cornet, all their arms and guns, and a large number of stock. On our way back to camp we burnt a*

> *few waggons and a house, and the Boers were sniping at us all the way... We rested today at a nice camping ground with plenty of feed, wood and water. The horses and mules are in good condition. Our black scouts went out this evening to find out where a commando of Boers are camped. There are many Boers here, you can see them all around on the skyline... The black scouts returned at sunrise and reported a Boer camp about two miles away... Mounted troops left camp last night at ten o'clock and marched in silence till daybreak when we surrounded a farmhouse, but the enemy had flown during the night.*

Trooper William Walder's diary consists of daily bulletins about snipings, skirmishes and endless marches. It is a seemingly-endless chronicle of events which depict two great white races of South Africa locked in a desperate conflict to the death; and in all this frightful business the Australians proved that when it came to a dogged shooting match they could stand punishment longer than their enemies.

> **August, 1901.** *Our convoy did not move today, but the Flying Column went out and engaged a commando of Boers. The 18th. Hussars lost three men and the Boers had three killed... Mounted troops left camp at ten o'clock and marched in silence till daybreak and then the West Australians charged a farmhouse, capturing twenty-five Boers at the point of the bayonet and their equipment, horses and two Cape carts. Later, the Scottish Horse charged a kraal and captured four Boers belonging to*

> General Boths's staff, also a complete heliograph machine (this is a signalling apparatus that reflects sunshine for the purpose of sending messages across a distance).

At this stage of the war the enemy's tactics were proving very successful as their marauding bands mounted 'hit-and-run' attacks on slow-moving British columns; and it seemed as if the Boers enjoyed blowing up railway lines and bridges, causing many British soldiers to lose their lives. Consequently, Lord Kitchener launched a "scorched-earth" policy against the Boer guerrillas aimed at depriving them of any horses, food, clothes and shelter that might be supplied by sympathetic homesteaders.

> We arrived in Ermelo about two o'clock and the Boers left the town as we entered it. We destroyed the town and all furniture in the houses, including about forty beautiful pianos and organs — thousands of pounds of furniture went up in smoke... We left Ermelo this morning and came on to Haalfontein. We had three men wounded today by snipers, one was a corporal in the Scottish Horse and two were Mounted Infantry... We brought the convoy to Carolina, a nice little town, or rather the ruins of one. The Boers were sniping at us all the way but did no harm. We brought in women and children, captured stock, and about twenty prisoners-of-war. Many troopers are dismounted, leading their weak and dying horses... The weather is fine now. About four hundred Mounted troops with one Pom-pom went out today and brought back a dozen families, cattle, horses and sheep... We were packed up this morning ready for

another march, but a very heavy fog descended and we were ordered, "As you were!" which means off saddle and pitch camp again... Two troopers of the Mounted Infantry were sent out on foot and armed with the white flag to spread Lord Kitchener's proclamation that "after the 14th. September, 1901, all Boers captured in the field shall be transported from South Africa for life." The Boers simply laughed at it, but treated the messengers very well then sent them back to the column to tell our Commandant that they would not come in till caught.

And so the Boer resistance continued — and so did the suffering of innocent women and children in the refugee camps.

August, 1901. *We struck camp this morning at seven and marched about nine miles over hilly country towards Middle Kraal. We brought the Carolina Town Guard with us after blowing up the old fort there. We had to leave behind twenty of the Scottish Horse who are down with Enteric Fever... The Boers attacked our rear ~ guard heavily, killing one man of the Northumberland Fusiliers and wounding three others. We camped in a valley near a running stream, but the poisonous tulip is very plentiful here and a great many of our horses are dying through eating it. We stopped at Swift River Camp and a large number of troops indulged in a bath this evening, the first I have had for eight weeks on account of the continual marches and dangerous country. We all had to sleep with full equipment on and our loaded rifles beside us, with two hundred men doing picket-duty every night to guard against* *possible Boer attacks... We struck camp this morning at six o'clock and started marching towards Middle Kraal, but the Boers commenced a rearguard action and by 12 o'clock it was so heavy that the convoy had to be laagered (waggons drawn up in a circle) and all available men sent back to reinforce the rearguard. Several men of the Eighteenth Mounted Infantry were wounded very badly.*

The object of Colonel Benson's column was to seize the Bothasberg Hills in Northern Transvaal and flush the Boer commandos out of their mountain fortresses. So skilful had Benson and his men become at night attacks in this difficult country that out of twenty-eight attempts twenty-one resulted in complete success, and Trooper William Walder's diary proves he played a brave part in the success of Benson's operations. (**Eventually, the Boers made Benson pay dearly for all he had done against them.**)

September, 1901. *We arrived at Middelburg after a long trek of six weeks, and left the Highlanders at Witbank to guard the coal mines. The Boers were continually sniping at us at long range. The weather is getting much warmer now. We go into town whenever we like, but we have to leave before six in the evening because of a curfew... We left camp this evening and after a night march of about thirty miles we surrounded a small commando of Boers at daybreak and captured them and their convoy, including six hundred head of cattle... I started working as a Farrier for one shilling extra per day... An empty convoy taking all the prisoners,* stock and families left here today for Wonderfontein and will bring back provisions for our column.

Night-marches usually had to be done on very short rations for there was no food to be found on the open veldt and as one soldier said, "A crow must carry his own rations when flying across it."

> **September, 1901**. *Mounted troops left camp at five o'clock last evening and marched all night until eight o'clock this morning when we charged a Boer laager, capturing fifty-four prisoners, the whole of their convoy of fifty waggons and carts, fifteen hundred head of cattle and a number of horses and sheep... Our column travelled twelve miles today through heavy rain, and some of the flanking troops lost the column, unable to see it through the mist and rain. We arrived at Carolina and went round the Kaffirs' kraals looking for mealies (African millet) and we got two waggon loads. I have been kept busy shoeing the horses... My "H" Squadron is under the command of Major Murray and I am getting the horses all shod up ready for a night march... We marched about thirty miles last night and at daybreak charged a small laager capturing thirteen prisoners, cattle and horses. The prisoners were in a terrible state, wearing rags, no boots, and starving.*

In this latter part of the war, the Boers lacked food, clothes and ammunition so they often "borrowed" the uniform of any soldier that fell into their hands and then sent the poor fellow, naked and blushing, back to his camp.

October, 1901. *We marched all night. It was very dark and cloudy and at daybreak we charged about three miles at a gallop into a Boer laager. They saw us coming and tried to escape but we galloped after them, capturing six and killing one. We also got a waggon and some Cape carts. We then returned to the column and later found that one man was missing from our Scottish Horse, but he had only fallen off his horse so turned up later. Orders received today said that we would not be told anything about any more night marches until an hour before starting. This is to avoid any spies in our camp warning the Boers we are coming. Too many times the Boers have flown just before we reached their laager and the chief has decided it is not just coincidence. The Boers fired into our camp last night, killing one man in his bed. He was buried this afternoon.*

It was the beginning of Spring when Trooper William Walder reached the town of Middelburg, situated on the Delagoa Railway. Both the men and horses were not sorry to have a rest in camp after five weeks of constant trekking.

On the 12th. October there was a farewell concert for the Argyll and Sutherland Highlanders who had served in Colonel Benson's column. They were leaving to spend the next three months on duty in blockhouses along the line of communications, a restful change from constantly riding across rough country on short rations in foul weather.

On October 15th. Trooper William Walder's time with the Scottish Horse in Colonel Benson's column was at an end, so he left Middelburg on a train for Johannesburg to obtain his discharge. It was in this city that Trooper Walder and his comrades met the Marquis of Tullabardine who thanked them for their fine contribution to his regiment, praised them for their courage, and offered them the honour of serving in his famous Scottish Horse for a further six months. It seems, however, that William refused the Marquis' generous offer.

He greatly admired Johannesburg and almost decided to stay there with his two chums and join the city's Police Force. Instead, on the 25th. of October after a comfortable train journey, William arrived back in dusty Cape Town where he eventually received his discharge from the Scottish Horse and was able to discard his khaki uniform at last. He later went down to the wharf to bid farewell to his chums who were sailing home to Scotland.

While lodged in Cape Town looking for a job, William heard the sad news about the fate of Colonel Benson and his column. Apparently, at the end of October — just two weeks after William had left the regiment — Benson's column was moving towards Brakenlaagte in a deluge of cold driving rain when it was furiously attacked by a combined force of Boer commandos led by Grobler, Oppermann and Botha. They took their revenge upon Benson, a just retribution for all the trouble he had caused them in the past months; and the battle-toll for the British was 60 men killed and 170 men wounded.

It must have been a terrible shock for William when he heard that sixty-two of his comrades in the Scottish Horse were dead or wounded, and that three of his former officers — Colonel Benson, Major Murray and Captain Lindsay — were also dead. That great soldier, Colonel Benson, was shot in the knee and stomach and as he lay dying uttered these last words, *"No more night marches".*

PROFILE – WILLIAM RUSSELL WALDER

Walder joined the Cape Town Railways as a porter and then trained to be a signalman. He stayed in South Africa until the following May when he decided to return home. He arrived in Melbourne on 6th. June, 1902, and returned to Watchem where he took up farming and made a success of that occupation. He married Miss Ellen Shepard of St. Arnaud and they had three sons and two daughters. William was a faithful and active member of the Church of England in Watchem where he worshipped for many years; and with true Christian charity he always supported any worthwhile movements for the benefit of Watchem district and residents. He was a member of the Burnaby Masonic Lodge for sixty years and a member of the Watchem Cemetery Trust. Certainly he was a man imbued with a fine community spirit. Four of his brothers fought in the First World War and one of them was killed in France. William's son, Harry Percival Walder, served his King and country in the Second World War. When William retired from farming in 1955, he left Watchem to live in Kerang where his wife died in 1965. William passed away at the age of 88, on 6th. June, 1973, exactly seventy-one years after he arrived back from the Boer War.

Ian Walder, of Bendigo, told me that his grandfather, William Walder, was a very proficient wood-chopper from an early age; and by the time he was only 15 years old he was physically strong enough to be the 'Pacemaker' in wood-chopping competitions around the countryside.

THE SCOTTISH HORSE"

When you read about the Gordons and the glories of the Greys,
The Black Watch and the Seaforths, and you sing in Scotland's praise;
Don't forget to Scotland's glory there is still another force
That will live in Scotland's story — it's the famous "Scottish Horse".
They were only ranked as Yeomen, but they quickly made their name,
Feared and honoured by the foemen and now on the scroll of fame;
Stands Tullibardine's heroes, auld Scotland's Yeomen force,
The equal of her glorious Greys — the famous "Scottish Horse".
When a bustling Boer commando rides the veldt on mischief bent,
Boasting that to wipe out "Khakis" is their one and sole intent;
If they hear the "Jocks" are coming then they quickly change their course,
They've a sad and sore remembrance of the famous "Scottish Horse".
Colonel Benson's fight near Bethel, where explosive bullets fell,
And the veldt, blood-red and filled with dead, seemed like a living Hell;
There was no thought of giving in to that overwhelming force,
For "surrender" is a word unknown to the gallant "Scottish Horse".
In the dim and distant future when the years have rolled away,
And your prattling, toddling babies have grown older, worn and grey;
They will tell their children tales from this sad and warlike source,
But by far the best true stories are about the "Scottish Horse".

This poem was brought back from South Africa in 1902 by

Trooper William Russell Walder and written by one of his chums in the 2nd. Scottish Horse. It immortalizes the outstanding courage of their Scottish Horse Regiment by describing the fight at Brakenlaagte, near Bethel, which took place on the 30th. of October, 1901, when a body of the 2nd. Scottish Horse and a number of the Yorkshire Mounted Infantry were assisting Colonel Benson's column in gathering up Boer prisoners.

Suddenly, through the heavy curtain of mist behind them, a trooper saw enemy horsemen bearing down upon their rear. *"There's miles of them!"* he yelled as a warning. Like the waves of a mighty ocean, several hundred horsemen burst through the thick curtain of mist, charging towards the British guns; but the Scottish Horsemen swiftly dismounted to form a protective human screen around the guns.

Then the Boers, stopping to form one long line, simply poured a withering hail of bullets into those brave fellows who fought back all attempts to capture the British guns. It was a brutal fight and the casualty figures speak for themselves for there were 62 killed or wounded out of 80 Scottish Horse, whilst the Yorkshires were practically annihilated. Yet not one man flinched or fell back to create a gap in the protective human ring they had formed around their guns.

Colonel Benson was hit in the leg and stomach and died next morning uttering these last words, *"No more night marches!"*

Only a soldier who has seen with his own eyes the aftermath of such a battle could write the following words: -

"The veldt, blood-red and filled with dead, was like a Living Hell".

Troopers Richard Merrett and William Russell Walder (both born at Watchem) had joined the Scottish Horse so they probably knew DANIEL JOSEPH GRANT who was also serving in the Scottish Horse and was wounded twice. Merrett and Walder must surely have met Grant again after the war when he was a policeman stationed in Donald. Daniel Grant died tragically in 1905, aged 26 years, by shooting himself in his bedroom at the Donald Police Camp. A suicide note stated his life had been made unbearable by another officer in Donald.

Mounted riflemen, like Troopers Merrett, Walder and Grant, had a special mission to perform. Often they were sent out on patrol with instructions to find and attack some hidden Boer camp: -

"When the night is dark and stormy winds blow
'neath rolling clouds hung black and low,
The brave patrol brings forth his steed and quits the camp with silent speed.
No bugle call or roll of drum bids him "God speed!" The camp is dumb.
And on the way no comrade's song or merry jest cheers him along.
Silently marching goes the brave patrol, forth to duty's call,
Fearing not to fall. With straining eye he forward goes
To find the lair of ambush'd foes; and oft he holds his panting breath
When face to face with lurking death. The foe's awake, he hears their cry
And lifts his voice in brave reply. Hill tells to hill the echoing news,
While foemen fly and he pursues."
(by A. Horspool)

This contemporary cartoon illustrates more vividly than any words can, how tiresome and frustrating it must have been for our soldiers as they chased the enemy from end to end of South Africa. As Tommy Atkins said: "I came out here to fight, not for a bloomin' foot-race." The frequency with which the Boer was cornered, only to escape, became decidedly wearisome and monotonous for the British Lion.

"THE SCOTTISH HORSE"

Men may come and men may go,
But I go on forever.

PROFILE – JAMES MUIR

Muir experienced some very tough times while serving with that famous regiment, the Scottish Horse. It was a dangerous pastime, chasing and fighting roving rebels across the plains and kopjes of South Africa.

One story about James Muir is sufficient to illustrate his prowess as a Mounted Rifleman; and that story came from Police Constable, William Morrison, who was stationed in Donald from 1924 -29. Morrison was a Boer War veteran who had served with James Muir in the famous Scottish Horse. He said that they were in a charge one day against the Boers when James's horse was hit and both animal and rider crashed into the midst of a seething melee of men and horses. He watched in horror as James disappeared from view and he thought he would never see James alive again. Amazingly, Trooper Muir emerged from the fight, uninjured and untroubled except for the minor inconvenience of extricating himself from under his dead horse.

James Muir was officially welcomed back to Donald on 19th. November, 1901. He married Harriet Mitchell and they had three children, but the year 1902 was a terrible year of drought and duststorms bringing ruin to many local farmers. Not surprisingly, James Muir took his family away to New Zealand where farming prospects looked much better.

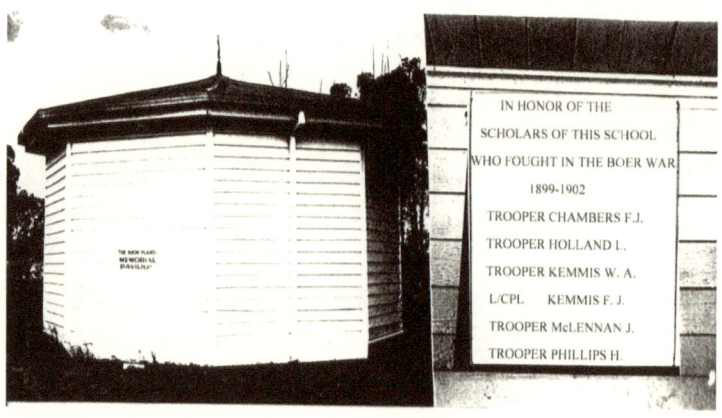

Photos. Ron Falla Collection

A BLOCK HOUSE WITH ITS PROTECTING WIRE-FENCE

In October, 1900, President Kruger left the Transvaal, abandoned his people to their fate, and escaped to Holland taking his country's money and archival documents with him, just like a common thief in the night. From that moment the Boer position was considered to be hopeless. Still, a few of the Boer leaders such as De Wet, Botha, Smuts, Viljoen and Delarey, decided to carry on the resistance by using guerrilla methods of warfare.

At first the British thought it would be a simple military operation to hunt down and capture these small bands of rebels; in practice nothing could have been more difficult and this period of guerrilla warfare was amazingly successful for the Boers. Our slow-moving British columns with their heavy guns and supply-waggons found it an impossible task to catch fast-moving Boer commandos who were familiar with the countryside.

The war dragged on and more lives were lost. Consequently, Lord Kitchener was forced to become increasingly harsh. He had the brilliant idea of constructing a network of guard-posts known as 'blockhouses' to stretch across the countryside close to the railway lines. These chains of bullet-proof, iron 'blockhouses' were linked by telephone and connected by miles and miles of barbed-wire entanglements. Thus the Boers' freedom was restricted and the British columns were able to sweep the countryside, driving the commandos against the blockhouse lines and into their net.

LIFE ON BOARD A TROOPSHIP

ALONG THE WAY

The S. S. "VICTORIAN" is carrying quite a large cargo of livestock. The ship has 4 officers and 86 crew, including 13 stowaways. The contingent comprises 34 officers and 596 men. There are 778 horses, 4 dogs, 3 cats, 1 monkey, 2 cockatoos, 1 canary and numerous rats. In the history of the world no ship has ever sailed such a long distance with such a large cargo of horses. Let anyone disprove this who can!

On the 5th. May one of our horses died from enteritis. As usual a fatigue party was told off to get spades and bury the dead horse. Five or six men promptly proceeded to obey, forgetting that they were not in Langwarrin Camp. It was not until the vet saw their spades and was heard to say, "None o' yer jokin' aboot this job," that the situation dawned upon them and the horse was dropped into the sea.

Our three doctors have enjoyed themselves immensely; with fiendish delight they have vaccinated the whole contingent. The command "Present Arms!" has taken on quite a different meaning.

The 778 horses consume nearly 4 tons of hay and 3 tons of bran each day; and each day the men consume:- Beef/mutton=1500lbs.   Flour=800lbs.   Potatoes=2tons.

Butter=150lbs. Biscuits=7lbs. Sugar=80lbs. Rice=40lbs. Tea=12lbs. Coffee=16lbs. Fresh fish=90lbs. In addition to this there is an enormous consumption of cordials, jams, tinned meat and fish, pickles, sauces and confectionaries which the men purchase with their own money.

Many of the men do not like the shipboard tea as they are used to billy-tea, and they complain that the ship's tea does not have any taste.

AN ODE TO LANGWARRIN MILITARY CAMP
Farewell, Langwarrin, a wanderer bids thee
A lasting farewell to thy damp dismal plain;
And his loving, last wish as disgusted he leaves thee
Is that his eyes may never behold thee again!

EXTRACTS FROM A SHIPBOARD DIARY

May 1st. (5 p.m.) Embarked on steamer after numerous farewells — kissed one girl by mistake and felt sorry, but she looked pleased — vow I will kiss no more girls till landing at Biera — reflected on congratulatory, farewell speeches of my friends — meditated on my courage at volunteering and my pluck in braving the hardships of a military camp — my chest swelled visibly — felt patriotic.

May 2nd. (3 p.m.) It must have been the soup — constant churning of internal arrangements — explored the bulwarks — kept a firm grip of the chains for an hour — eyes right — saw several comrades on similar duty — wonderful fascination of the sea directly in front of us — eventually lay on deck — heard bugle call for tea — wasn't having any!

May 3rd. Wasn't hungry, thirsty, sleepy, happy, or anything.

May 4th. Still afloat — found others eating everywhere, the consummate gluttons — here was I for two days and had nothing — surprising pluck — if only the officers knew (they probably did).

May 5th. Roll-call so managed to reverse my usual position on bulwarks — announcement of appointments — now I would hear something of interest — I remembered with great satisfaction my camp experiences of 30 days (the 6 days' drill, 14 days drawing kit at the Langwarrin Railway Station and 10 days on other duties) — knew right hand from the left hand, but a bit uncertain about my feet — happy thought, I would chalk my left boot and keep my eye on it — surely I would hear my name as Sergeant — no — strange — ah, well, one would be satisfied with two stripes — foiled again, not even Lance/Corporal — probably overlooked — when I got better — but on second thoughts I turned right about and looked into the water with a long, steady, deliberate gaze and repeated several former extension exercises.

May 6th. Saw no whales, no birds, no sharks, no ships — didn't care. Heard that we had left Australia behind — felt sorry for Australia.

May 7th. Reclining on deck — wash or shave would be pleasant — too much trouble — dreamt of home, of Langwarrin, of Flemington, of the meal at Albert Park — awoke, felt hungry, ate two bottles of pickles.

A SOLDIER IN SOUTH AFRICA

John William NEYLAND, of Corack, sailed with the Fourth (Imperial) Contingent for South Africa on the transport "VICTORIAN". The ship left Port Melbourne on 1st. May, 1900, carrying a total of 631 men, 778 horses and 11 waggons. The men selected had to be seasoned bushmen, bold riders, sharpshooters and capable of contending with a guerrilla enemy; and when joining this contingent the men were given a choice of either serving for a period of twelve months, or staying for the duration of the war. From the troopship "VICTORIAN" Sergeant J. W. Neyland wrote this letter to his father: -

"*17/5/1900 — We are now the Victorian Battalion of the Australian Imperial Regiment and pride ourselves accordingly. Of course, the third stripe on my arm which gave me the title of sergeant some ten days ago is duly appreciated. An unparalleled voyage as far as weather conditions are concerned has been our lot, but for all that many officers and men now have a personal experience of the definition of Mal de Mer. But four days saw all in order with improved appetites.*

Monotony is not allowed to creep in. We all have

our share of duty, but not in the rough way that falls to the lot of the private. Whether it be stable, fatigue, guard, or any of the other multifarious duties on board, our Australian Tommy works with as good a grace as possible under the circumstances.

Volley-firing caused a great deal of contention between the companies yesterday. My company proved victorious, with E Company a good second. Their target was a fruit case, thrown overboard, the firing distance being about 200 yards.

This reminds me that the generous donors of all the fruit have received many cheers from the lads, with hearty enthusiasm, to say nothing of our gratitude to the firms who so graciously supplied us with tobacco and cigarettes for the long journey.

Our musical evenings are a source of pleasure thanks to the generosity of Victoria's Governor-General. His piano is placed on deck, protected by its tin case; his concertina is also popular; and the cornet can be heard all over the ship; whilst the tin-whistle is played to the pirouettes of the soldiers.

Who amongst us can ever forget the glorious 1st. of May on Port Melbourne pier? That sea of upturned, passionate faces, glowing with colour and pride at our young volunteers, is a scene which will live in our memories while camped out on the lonesome African veldt.

While sailing across this glassy ocean our memories turn to family and friends and we wonder what it will be like for us in the hour of danger with every nerve on the alert? Our Colonel is that typical stamp of an

Irishman whom officers and men would follow to the death. His daily lectures bristle with points of instruction, interspersed with that special wit that seems to belong to men from the Emerald Isle.

In a recent lecture he said, "There are those amongst you who do not know what FEAR is; but I have heard there are men here who are not possessed of sufficient intellect to understand the meaning of the word DANGER." Silence reigned whilst many men pondered over his meaning.

Have I been vaccinated? Yes, and we had to face two ordeals! Three doctors revelled in the butchery and over 600 arms are now as ripe as strawberries. Some there are who have had a really bad time of sickness, but I am thankful to say I proved to be one of the lucky ones.

Of the 778 horses aboard we have lost but one up-to-date, notwithstanding the fact that we carry twenty from the sick lines at Langwarrin.

We have with us youths from the back blocks of Australia who had never visited a city before, and as for the sea, well, they have only read about it. One of these innocent ones asked me, "As this troopship has not previously visited Beira, how the Devil then can the skipper tell the way he is going?" A suppressed laugh was my response.

What would we not give for news from the front! The suspense is aggravating and all of us are actively counting the days up to landing-time. "Will we be too late?" is the daily question. All on board are anxious for the arrival of the time when our serious business can begin. On

Sunday, the 20th. of May, all eyes were turned towards the "Promised Land" and we saw — MADAGASCAR!

After mid-day nothing was to be seen but the usual sea and sky. On board it was a day of events to keep us busy, and when orders were read out we learned with pleasure that we are now the leading company. This was the result of all the competitions involving shooting, discipline, appearance, stables and mess-tables.

It is Monday the 21st. of May and the winches are creaking, commanders are shouting, and all men are agog with excitement at the prospective landing tomorrow. Inspecting the men's kits and restoring order out of chaos has caused time to fly and all hands are busy with the final preparations for our eventful life. Time alone will show what stuff we are made of — and afterwards we look for a glad return to the land of our birth."

THE FAME OF THEIR COURAGE

"As soon as the Bushmen landed here, three British Generals met at Headquarters to see Lord Roberts. While they were waiting the first general said to the second one, "What are you after?" "Come to see about more supplies. And you?" "Same as you." Here the third general broke in with, "I expect you are both here for the same reason as I am — to try and get those new Australians, the Bushmen." Shame-facedly, they admitted it was true.

Sergeant John William NEYLAND of "A" Squadron (Fourth Contingent) arrived at Biera on 23rd. May, 1900. He writes from his camp at Marandellas on the 12th. of July: -

"The VICTORIAN lay at anchor off Biera for a week. We had to stay on board while the N.S.W. contingent landed their horses; then it took four days to land ours. It was indeed a tedious process as the animals had to be taken from their ocean prison by means of slings, and lowered on to lighters which then conveyed them ashore, a distance of half-a-mile — though anything like speedy delivery could not be reckoned upon owing to lack of pier accommodation.

The Port of Biera is predominantly a reception place for troops, horses and a thousand other necessaries pertaining to modern warfare; so much so that the narrow — gauge railway from Biera to Bamboo Creek (60 miles) has been changed to a broader gauge of no mean order. Biera itself is quite modern and would have immense possibilities, but for one word — MALARIA.

Still, the salaries paid here are large and love of gold has well-populated the place. After ten days' camping we were overjoyed to leave this port behind us, but sorry to entrain without our Colonel who is struck down with fever.

Bamboo Creek was our next halting place and proved to be a more pestilential place than Biera. Its nomenclature was derived from the number of Bamboo canes growing along the banks of its stream; and the place itself is full of canteens and business establishments because the railway traffic since the war has brought great trade to the town.

The countryside abounds with game, but we could have none of it to eat — only black flies with black dust was our portion!

After two days in Bamboo Creek we entrained for Umtali, 222 miles from the seaport of Biera. GLANDERS (a contagious horse disease with fever and swelling of glands beneath the jaw) was responsible for the loss of a number of our good horses before we left here. These animals were shot wholesale, their mangers and feed-bags burnt, and their paddocks placed in quarantine until the wave of the disease was stemmed.

As no coal exists in these parts, wood is used to heat the engine's boiler and the pace of our train rarely exceeds 12 miles an hour. To negotiate a hill was a difficult matter and the amount of burning sparks emitted from the engine would suffice to set all Australia on fire

Umtali is pleasantly situated between towering hills and contains about 3,000 white people. After three days' rest in Umtali, Captain Chalmers of my "A" Company decided we would march to Marandellas, a distance of 125 miles. Anyway, the railway line was choked with traffic owing to the great number of troops en route to the front.

We were accompanied on our march by a hundred of the 71st. Yeomanry who were on foot because their horses were not ready and besides the majority of them had never previously sat astride a horse. We accomplished the journey in ten days, arriving on 11th. July. It was a pleasant ride with grazing in abundance along the way.

Our heavy kits were packed inside a bullock waggon in charge of a Boer. This man's brother is at the front fighting against us and yet to use the man's own words, "I don't care how long this war lasts because transport work for the British means any amount of money for me!" His 'race hatred' may be smouldering, but 'filthy lucre' is predominant in his mind.

Here at Marandellas we are 3,000 feet higher than Biera. The spot is healthy — warm days but bitterly-cold at night, similar to our Victorian spring. Food supplies, men and horses are being pushed rapidly forward to Bulawayo, Salisbury and Victoria, with all speed by light mule waggons that are pulled by 12 animals while oxen pull the heavier materials.

A number of our lads have thrown in their lot with the Royal Artillery encamped here, and other Australians are training horses for the Imperial Yeomanry to ride. The remainder of our Victorian Battalion arrived yesterday by road from Umtali and we leave on Saturday for Bulawayo. We now get a chance to be in it before the finish."

Sergeant John William Neyland would soon experience the sudden and nasty dangers of guerrilla warfare as he travelled through the Transvaal. From Lichtenburg to Komatipoort, roving Boer commandos were everywhere waiting to pounce upon unsuspecting convoys and railway trains and for ever seeking opportunities to harass the British.

Sergeant Neyland arrived in Mafeking in time to join Lord Methuen's troops (which contained a large proportion of Australian bushmen) and from then onwards he fought in the British operations throughout a difficult and important district that lies between Rustenburg, Lichtenburg and Zeerust. During the month of August, Sergeant Neyland was involved in many skirmishes with the Boers.

"My last letter was from Marandellas and since then

the times have indeed been stirring. We had a splendid march with Lord Methuen to Bulawayo, and from there to Mafeking by train to refit; and after four or five days' stay at the scene of Baden-Powell's triumph, scouts brought in the intelligence that the enemy was within three miles of the town.

So under Brigadier-General Lord Erroll, my "A" squadron with other Australian units, and the artillery with various regiments (altogether numbering about 3,500 strong) marched to the front with a convoy consisting of ox and mule waggons and two miles long. It was the panoply of war, an awe-inspiring sight to we Australians.

On the road to Otto's Hoop our scouts were fired upon, but it was only a small party of marauders who thinking discretion was the better part of valour soon disappeared back into the hills. Anyway, our main purpose was to hunt down De Wet and scatter the main Boer army under Botha.

We bivouacked in a Boer settlement that night and were placed under General Sir Frederick Carrington's command. He is a fierce old soldier with a large experience of South African warfare. Time was not allowed to hang heavily on our hands and the 16th. August, 1900, will ever live in our hearts as the date on which we received our 'baptism of fire'.

We left at a gallop for the scene of the action near Buffel's Hoek and arrived just in time to see the Boers leaving their snug refuge on a kopje in a mighty hurry. Our guns were drawn up and unlimbered and soon the Maxims and Pom Poms played havoc upon their broken ranks as they retreated.

With the Kimberley Mounted Corps and units from N.S.W. and New Zealand, our little band under Lieutenant Gilpin stormed and occupied the first kopje and from thence we progressed to the second and third kopjes while continual fire rained down upon us. Our first engagement lasted for ten hours and then Boer and Briton rested on their oars while darkness spread over the field.

A hard biscuit and the contents of our water bottles had to suffice the inner man, and then a few hours sleep in our overcoats (out on the open veldt) served to rest our weary bodies. During that skirmish, Captains Fuller and Harvey (New Zealand) and Trooper Gibson (N.S.W.) were shot dead, but the enemy's loss was heavy. My squadron's only casualty was one wounded horse.

The 20th. August ushered in our second tussle with the cunning Boer and a red-hot eight hours we had of it. We advanced in extended order to within 1300 yards of the enemy's trenches which were on a kopje above us, and they saluted us with a hail of lead. Not a sign of a Boer could we see, yet Botha and his merry men had our range to perfection. So we retired singly to the shelter of some bushes and waited for our reinforcements to arrive and then we were at it again hammer and tongs.

With Paget's Horse (they are brave men at holding their ground, but foolish at showing themselves to the foe) we poured in volley after volley to which Mr. Boer replied with interest. Bandoliers and pouches were emptied so more ammunition was drawn up; lead poured down on us like hailstones from above; and then the Boers' Maxim gun entered into the spirit of the game.

> They were lying at the top of a kopje waiting for our usual storming party, but we were not sending any on this occasion. How we wished for our own artillery to arrive and, fortunately, we did not have long to wait. CRASH! BANG! Heavy guns saluted each other and soon the darkness gave both sides a much-needed rest. Our orders were to hold the position and this we gamely did; but our roll call was a sad one.
>
> Lieutenant Gilpin, dearly-beloved by officers and men and ever foremost in the fray, was no more after his life's blood ebbed away from three Mauser wounds. Trooper Woodman (of Bairnsdale) was shot dead at his post in the firing line. An estimate of the enemy's loss is impossible, but truly the narrow escapes from death were marvellous and not a single wounded man from my own squadron was reported. Our soldiers were accorded a military burial in the little cemetery here known as "God's Acre" and their bodies lie close together in two graves at the top right-hand corner.
>
> Lord Kitchener has ordered the burning of all farmhouses from which shots have been fired at our men."

The following contemporary cartoon from an English newspaper sums up the feeling of most Britons who were suffering from this seemingly never-ending war. During the winter months of 1900 it became apparent to most people that Great Britain was now involved in prolonged guerrilla fighting. Lord Roberts' merciful treatment towards many of their countrymen had failed to persuade the Boer leaders to seek peace, so he had to adopt harsher measures to put down the fighting. Therefore in September, Lord Kitchener issued a proclamation to the effect that farms in the vicinity of train-wrecking episodes should be burnt — and thus began a "scorched-earth" policy.

18/9/1900 — *"We are at MALMANI, upon the Bechuanaland border and the night of 29th. August was the third time since landing at Beira on which Heaven's artillery has come down upon us — a different sort of bombardment from the usual Boer efforts. Pebbly hailstones fell and drenched us to the skin and caused a stampede of the horses in camp. Sleep was out of the question, but at daybreak order was restored out of chaos with the inner man warmed by rum and coffee and the outer man dried by the sun's rays.*

Our camp was quiet until 3rd. September — with the exception of continual sniping, deadly in some instances.

On the 3rd. of September, my "A" squadron of the A.I.R. acted as escort to a convoy proceeding to Mafeking for provisions. Nine miles along the track and a skirmish took place. We experienced the hottest fifteen minutes we ever knew, or ever wished for, in our lives. There was no cover nearby and so we sustained the full force of the enemy's bullets from about 400 yards. We had three men killed and ten horses shot.

Still, I managed to cover one Boer with my Lee-Metford and sent him the contents as a "billet-doux". Then we had to return to camp with our empty waggons.

On Sunday, 9th. September, a big movement was planned by Lord Methuen after re-equipping his forces at Mafeking. Methuen took the road to Lichtenburg; Lord Douglas moved East; and we Australians with

Paget's Horse, under Lord Errol, moved off at dawn to meet them. The plan was to encircle the Boers and catch them in our net.

We arrived in time to hear Lord Douglas's guns at work. His men had attacked a laager and accounted for 47 of the enemy in one stroke with his lyddite while they were sleeping; and we returned to camp with a score of prisoners who had been taken unawares and surrounded by my squadron. They may have been a tattered-looking crew in rags and worn-out boots, but they were able to use their deadly Mausers for all that.

By the 12th. September we had left Otto's Hoop and were camped with the South Australians, Tasmanians and a New Zealand battery. About three miles from where we had struck camp we could see the enemy clearly moving about on top of the kopjes; and during the afternoon a South Australian captain was shot dead by one of their snipers. Sniping is a daily occurrence and a vigilant watch must be kept up at all times."

Colonial troops were found to be so well-adapted by nature and training to the work in South Africa that the British military authorities showed a lively appreciation of their services by asking for more men of the same calibre to be sent out from Australia to replace those who had served their time and were returning home. As for a public and well-deserved recognition of their worth, one war correspondent wrote in an English newspaper, "The deeds of the Colonial forces in South Africa have done more to spread throughout the British dominions a strong patriotic feeling than all the politicians and their talk could accomplish in ages".

"Richtenburg Camp, 14/11/1900. We are in this district where Lord Methuen is operating with his forces. The town is an ideal country village and well-irrigated so that the crops, fruit, hedges and willows thrive. It has well-laid out streets and an abundance of greenery which is refreshing to feast one's eyes upon after months of weary travelling across the desolate veldt. Here is situated the homestead of the infamous General Delarey.

Since my last letter we have been working with Lord Methuen; and after we left Buffel's Hoek (near Otto's Hoop) we had to fight our way to Kaffirs Kraal and then on to the relief of Jacobsdal where the Cape Town Highlanders suffered a heavy loss. Their garrison of about 60 men had been ambushed by the Boers and two-thirds of that little force were ruthlessly killed, or wounded. Of course, these murderous Boers vanished as soon as we appeared.

However, we caught up with the enemy who had taken up a position on a wooded kopje, but our firing was so strong that they soon evacuated that kopje. Then the chase was on and "A" and "B" squadrons of the A.I.R. arose to the occasion. Never since our arrival here have I seen the Boers make such a race for their lives, although the hard gallop told severely on our horses and we could not match their fresh Cape ponies.

From kopje to kopje and across the open plain their retreat ended in a rout, until finally they took advantage of the mountains and found a hiding-place; but their loss was great and we returned to Jacobsdal with 30 prisoners and plenty of cattle, sheep and goats.

Our leaders are now becoming more cunning since it

> has been proved that any move made by our forces is known through spies to the enemy. So, now we pretend we are going to a certain place and then at the last moment we are led elsewhere.
>
> For example, on the 8th. of November we left camp at 4-30 p.m. presumably going towards General Botha's farm, but after leaving camp we turned eastwards away from our reported direction. You should have seen the look on our "loyal" Boers' faces — they had fallen into our trap."

After the British moved northwards across the Vaal River and took Johannesburg and Pretoria it looked as though the war was over; that was until a few Boer leaders decided to embark upon guerrilla warfare by using small mobile groups of commandos to mount "hit-and-run attacks" on the British — but always withdrawing before the British had time to respond. Sergeant John William Neyland described how he was involved in chasing the Boer rebels.

> "Taungs, Bechuanaland, 1/2/1901. We are still attached to the column of that hard-working General Lord Methuen, whose force consists mainly of Australian Bushmen and English Yeomanry. We left Western Transvaal just a month ago and travelled by train from Mafeking to Vryburg with our horses, mules and waggons. After three days of preparation at Vryburg we set out on a fortnight's trek to clear the disturbed parts of Bechuanaland.
>
> During those fourteen days we covered 280 miles, 60 of which we performed in two days. De Villiers and his men were very active, so chasing his commando we rode

40 miles one day and succeeded in driving the enemy out of Griqualand and across the railway-line close to Taungs.

But at a spot ten miles from Taungs the Boers blew up the railway-line in two places to gain time in getting away from us. During that long trek of 280 miles we relieved the besieged garrison at Daniel's Kuit where the small force had been surrounded by Boers pouring shells daily for a whole week into their camp until we arrived and made our presence known. By the middle of January we reached Kuruman and then marched to Schweizer-Renecke, clearing the enemy away from the borders of Griqualand.

Lord Methurn received great praise from the Commander-in-Chief for his long march and all the good work we had done.

Here at Taungs, which is a British garrison, resides a powerful native chief named Maluma whose sympathies are British to the backbone. In fact, at the beginning of this war he went so far as to offer to Great Britain 10,000 of his mounted warriors as fighting allies. Of course, his handsome offer was gratefully declined because this is a "White Man's War".

I might tell you that Taungs and district boasts a population of 32,000 natives. There is a decent railway station where I inspected the engine of a train recently ambushed by the Boers and I counted no less than 62 bullet holes in its casing. The train had been carrying provisions from Lord Methuen's stores for a nearby garrison. As the enemy was unable to steal them away they made a conflagration of them instead.

This new township of Taungs is proud to have a postal centre, telegraph and money-order office, two large hotels, numerous stores and comfortable private dwellings. It is situated on the Little Hart River and we have been camped on its pleasant banks since last Friday. A week's spell here has put new life into man and horse.

I learn that tomorrow morning at daybreak we are once again on the move. Still, our treks are never devoid of interest."

The Australian Bushmen continued to do good work with Lord Methuen's column. They left Taungs on February 2nd. and were fighting in skirmishes at Uitval's Kop, Paardefontein and Lilliefontein, in each of which the enemy was successfully brushed aside. Passing through Wolmaranstad, Lord Methuen's column turned north towards Haartebeestefontein. Sergeant J. W. Neyland continues his story: -

"**With the A.I.R. at Witmoss Railway Station, 10/4/1901**. *We have had a busy time. On the 18th. February our spies informed Lord Methuen that the Boers had left their laager near Haartebeestefontein to fight elsewhere, so we pounced at dawn and captured 10,000 head of cattle, 43 waggons and 40 prisoners. Next day we attacked De Villiers and his merry men and after five hours of hard fighting drove them back from the pass which they were holding against us. Our Australians and English Yeomanry did very well considering we only numbered about 1,500 and were fighting a larger force entrenched in a very strong position on top of the hills.*

Our British casualties amounted to 16 killed and 34 wounded, while the Boers left 18 of their dead upon the kopjes which they had abandoned.

We returned to Klerksdorp for a spell and then Lord Methuen took us westwards until we reached Warrenton on March 14th. There we parted company with Lord Methuen and journeyed by train for three days to Graaf-Reinet, a station on the Port Elizabeth railway-line and also one of the chief centres of Boer activities in Cape Colony.

From Graaf-Reinet we rode over arid, hilly country through a village named Pearston and on to Middlewater where we rejoined our "C", "D" and "E" squadrons of the A.I.R. whom we last saw at Marandellas in June. Here our time was filled with doing outpost duty on the tops of high hills and patrolling through the various gulches in the mountains looking for Boers.

After three of four days of hectic activity our column was ordered to move on after word arrived that the rebels had vacated this region of hills covered with prickly pear bushes. But while we were acting as the advance party to our column, Sergeant Hurst and Trooper Goebel disturbed a small band of the enemy and the poor fellows were shot dead by snipers in the impenetrable bush.

We marched back through Pearston and on to Thorn Grove, a distance of 45 miles which we negotiated in just two days. This land is barren of grass, but the local sheep and cattle seem to thrive on a certain plant known as the "Karroo bush" which somehow manages to grow even during the dry, waterless seasons and our horses greedily ate it for want of something more delectable.

From Thorn Grove my squadron went out on patrol-duty and came upon the rear-guard of some 400 rebels just as they were leaving a farm known as "Le Clereqs". We placed our horses under cover and by aid of the heliograph sent a message back to the main camp.

We were eating our dry biscuits for breakfast, sitting amidst the dense prickly pear bushes, when the sound of the enemy's Mausers drove all other thoughts from our minds and soon we were blazing away at anything and everything that appeared in sight beyond the dense, thorny shrubs.

We wounded a score or so of the rebels and one was killed, but during their retreat we noticed four armed kaffirs (natives) riding with them. My squadron had a single casualty, yet he was a very lucky man. Trooper J. Burke was hit in the chest, but his leather bandoleer turned the course of the enemy's bullet so that it passed through his left arm instead of his heart.

While following this commando of Boers, we came to a nasty corner between two hills and as Captain O'Farrell and his squadron galloped forward, a hail of bullets descended upon them from the enemy who were safely ensconced behind rocks in a strong position on the mountain side.

My "A" squadron then faced "running the gauntlet" and the rebels volleyed into our ranks as we galloped between the hills. The attack lasted an hour or more and then the rebels fled, their object of having delayed our advance having been achieved. We passed onwards, our various squadrons choosing different tracks across the mountains.

That night we camped at the homestead of a Mr. Trollope after carefully surrounding his farm and the various outbuildings. But the owner proved to be a loyal resident and gave us information about the rebels.

Apparently, on the previous night the rebels had stolen most of his provisions, but this kind and loyal man gave us all he could spare and did all he could for the comfort of our men and horses.

Early next morning we left Mr. Trollope and moved out carefully expecting the enemy to be nearby, but after some cautious scouting we found the enemy had retreated back into the hills. Evidence of their movements was obvious by the looting of all the farms along their way.

We moved on to Witmoss where the column is now in camp. When we are out on night or day patrols in this inaccessible country, our work is both hard and decidedly dangerous with the constant fear of a bullet in one's back

At the end of our twelve months' service in South Africa (on 23rd. May, 1901) most of us want to be homeward bound. Every inducement is being held out to us to stay on in the regiment. For instance, troopers will be placed a step higher in rank; sergeants are promised commissions; and officers will receive a higher grade. But twelve months of this dangerous life is sufficient for any ordinary mortal, so I hope to see you all soon."

When the Fourth Imperial Contingent left South Africa on the 26th. June, 1901, Sergeant John William Neyland had intended to be with them. In fact the Donald newspaper announced in July, 1901, there would be a joyful welcome awaiting him on his arrival home "especially by the Corack folk among whom his domicile is situated". The newspaper editor also made an interesting observation:

"The sergeant is the only one of many recruits who volunteered from this portion of the Wimmera district, who gained distinguished promotion; and the probability is that if he had had the advantages of an education in the higher schools he would have advanced much higher in the ranks."

After his regiment left South Africa, Sergeant John William Neyland remained there to handle and train the many horses used in the South African Constabulary. His two brothers, Niven and James, also worked with these horses.

PROFILE: — JOHN WILLIAM NEYLAND was born on the 30th. June, 1876, the eldest in a family of thirteen children. Margaret (Niven) and John Pringle Neyland were early settlers at Corack East and their nine sons grew up to be excellent horsemen who rode their father's fine horses in many races around the North-Western district of Victoria. In 1900, John William Neyland volunteered to fight in the Great Boer War and joined the Fourth Imperial Contingent. Enlistment records describe him as "a single man, a grazier, and a member of the Church of England, 5ft.-7ins. tall with a chest measurement of 38 inches". When eventually he returned from South Africa, he married Margaret Cavanagh, of Watchem, at St. Columbas Church in Hawthorn. They farmed in the Corack East area, growing mainly wheat and sheep. They had two children, Glory and John Thomas. In 1914 at the beginning of the First World War, John William Neyland volunteered again

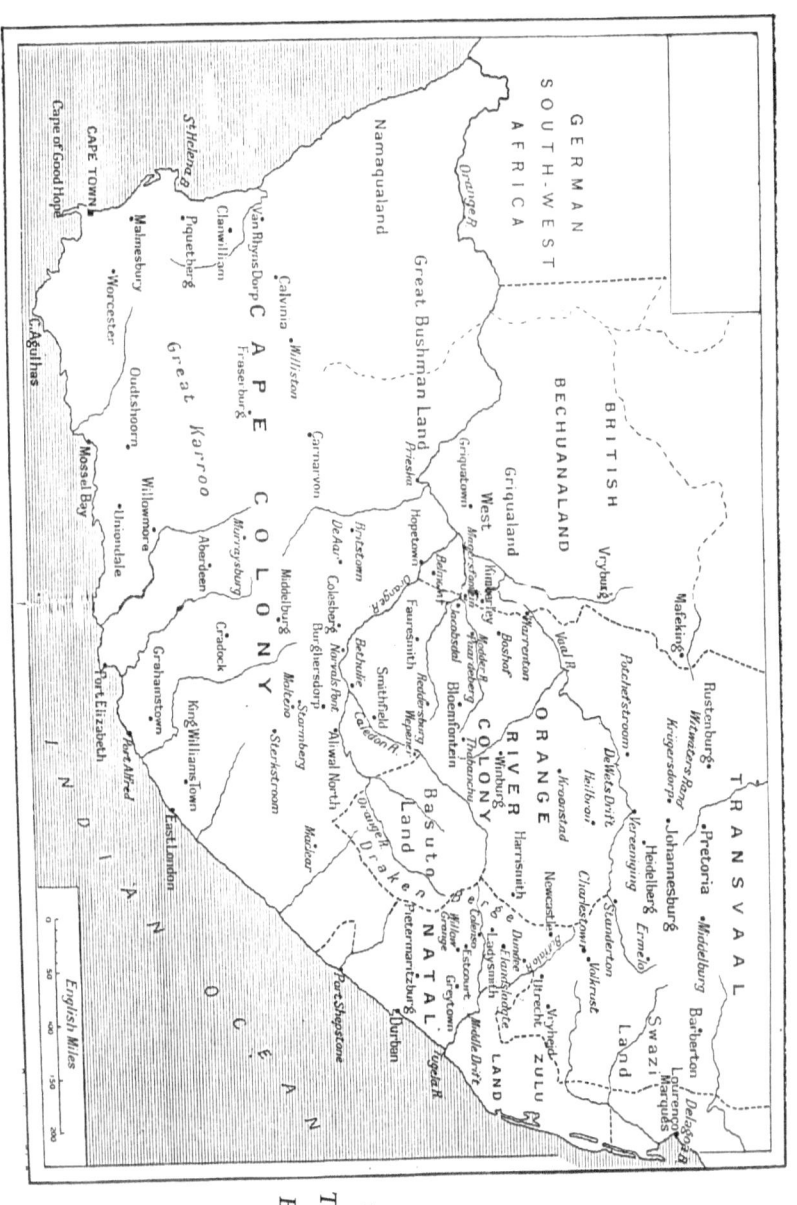

SKETCH MAP OF SOUTH AFRICA to illustrate THE GREAT BOER WAR.

On August 4, 1900, Major Tunbridge with 500 Queenslanders and New South Welshmen were surrounded and besieged by 3,000 superior armed Boers on the Elands River. After their horses had been killed, General De la Rey offered them a safe conduct if they would surrender. Tunbridge's reply was: *"If you want us come and get us."* Though abandoned by the relieving force the little band hung grimly on until rescued by Kitchener eleven days later.

to fight for his country — and so did all his brothers. In fact, nine Neyland brothers volunteered and eight were accepted, thus creating an amazing record for the highest number of soldiers from the same family fighting in the war. John William served with the Light Horse in Egypt, but was invalided home from the Middle East in October, 1916, and was then sent to New Guinea as an assistant to Governor Murray. One day during his patrol duty in the hills, he captured two German officers, survivors from the infamous "EMDEN" which had been sunk by H.M.A.S. "SYDNEY" off the Cocos Islands at the beginning of the war. One of the German officers was so pleased to be found that he gave John his pistol as a memento of the occasion. After the Great War, John William Neyland became an orchardist on the Murray River, near Swan Hill; and in 1939 it was the turn of his only son, John Thomas Neyland, to carry on his father's fight for Justice and Freedom. In 1946, John William Neyland died at Swan Hill, aged 70, a veteran of two wars.

N.B. John William Ncyland's daughter, Glory Allison (born in 1908 and now living in Bendigo) remembers her father as a wonderfully kind-hearted man, always going to the aid of any neighbours suffering hard times due to drought and rabbits.

DOCTOR NEWMAN'S STORY

The British occupation of Bethlehem (Orange Free State) occurred on 7th. July, 1900, and was vividly described in Trooper Walter Coombs' letter. Doctor Newman relates his experiences of the very same event, but puts the emphasis upon medical aspects of the battles. Sometimes it is difficult to know exactly where all our colonial soldiers were fighting because they were scattered among the British units.

> "**Bethlehem — 12th. July, 1900**. *We have had some very lively times since leaving Lindley. We left at 7 a.m. for Bethlehem in the south-east of the Orange River Colony. We expected to meet a good deal of resistance and indeed we did, for we were six days getting here and we fought five battles; and on the day on which there was no battle we suffered a lot of sniping from invisible enemies.*
>
> *Shortly after starting, about three miles out of Lindley, we saw a considerable number of Boers near Leeuw Kop and shelling began in earnest and continued all day. Our men went forward to outflank the enemy and force them to retreat, but as we advanced the Boers gave us a very warm time. We camped for the night at the foot of this kopje and their three guns were shelling us with great*

accuracy. The Boers were led by DE WET and our lot by GENERAL PAGET. Altogether we had 4,000 men in our column, including Imperial Bushmen from South and West Australia.

Next morning we moved off at 6 a.m. and had only moved about half-a-mile when their shelling began. The Boers made excellent hits with their long-range guns but fortunately no one was injured. After we moved on for about three miles, the Boers took up a position to the left of us and our infantry advanced to outflank them, so a great battle followed.

Our ambulance advanced with the infantry until shell dropped right in front of us so then the ambulance had to retire a little, but it was impossible to keep out of the range of their big guns. It is amazing though that so little damage is done by the shells.

We had one case of an Australian Bushman struck in the left knee. Unfortunately for him, the bullet drove a piece of clothing into the joint. I have operated on him twice, but I am afraid it will end in the loss of his leg.

Whilst we were fighting, General CLEMENTS' column arrived and began shelling the Boers. By about 2 p.m. our infantry got close enough to the enemy and there was a great fusillade which was kept up till dusk. Then it came on to rain in torrents and was bitterly-cold.

Our ambulances were sent for and my commanding officer and I went out to find the wounded. The Bearer Company informed us the wounded had been carried to a deserted farmhouse nearby and there we had a busy night's work. We were going at it for hours, dressing their wounds. There were two dead and eleven wounded (one has since died and two were beyond our help).

During the night our ambulance went out in search of more and we stored the wounded at another farmhouse. To cut a long story short, the Boers then retreated and our casualties amounted to twenty. By the way, our Australian Bushmen did well in the fight on July 3rd. by saving our guns which came within an ace of being captured. A small band of Australians came to the rescue with alacrity in the face of sharp rifle fire, and drove the Boers away from the guns.

That night we slept at the farm and next day removed all the wounded to another farm and left them in charge of one doctor. We had to move on with the column because during an advance we have to go with the fighting soldiers and leave the wounded behind.

On July 6th. we camped about four miles from Bethlehem. The place is surrounded by hills and the enemy was strongly posted. Several of our cavalry actually got into the town, but the Boers wounded several of them and killed one. There was really no necessity for them to advance into the town, but some men, especially the Yeomen, are brave but foolhardy, and do the most ridiculous and dangerous things.

The shelling of the town went on all day and we had four wounded and one killed. The Boers started shelling our camp with excellent accuracy, their shells bursting within a few hundred yards of our tents, so some of the infantry had to move back.

At 9 a.m. we sent in a flag of truce, demanding the surrender of Bethlehem into our hands, but the Boers refused this offer and the shelling continued. Our force

brought forward their two great guns and the rest of the day was pretty lively. These guns are tremendous things, about 15 feet long and drawn by oxen.

During the afternoon General PAGET, hoping to find the enemy's weak spot, sent forward the Royal Munster Fusiliers supported by some Australian Mounted Infantry to attack the Boers' left flank; but they fell heavily, having two killed and forty wounded, some of which have since died.

Apparently the Boers were hiding in a donga (gully) and as the Munsters were crossing a mealie field the Boers opened fire on them. The enemy's fusillade was tremendous, but finally our brave Munsters charged the enemy's position and took it.

That night I had plenty to do and we were wandering about on the veldt till 11 p.m. before we had collected all the wounded. The next day was busy as we were operating and dressing cases all day long.

The guns, big and small, began to sing at daybreak. Our two columns, the left under PAGET and the right under CLEMENTS were hard at it. I don't know how many guns were engaged, but between them there was a constant din. The boom-boom of the big guns and the click-bang of the small ones kept things lively until noon

Then General CLEMENTS sent forward the Royal Irish Brigade. Up they went in three extended lines, dropping many men on the way, but they reached the crest in time to see hundreds of the enemy horsemen retreating; and they were able to capture a Boer gun (15-pounder).

This Irish Regiment bore the brunt of the action and succeeded in forcing the Boers out of Bethlehem, but not without a heavy loss of sixty killed and wounded.

That evening (7th. July) General Clements and his column occupied Bethlehem. On Sunday we shifted our wounded into the town, turning the Town Hall into a hospital. We then commandeered all the beds, blankets, mattresses and utensils we could lay our hands on.

The town is rather pretty and the principal building is the church, as in all Dutch towns. The houses were empty, the people having charged out and left their places in great disorder; and needless to say the disorder was considerably increased when Tommy Atkins arrived to help himself. We commandeered a small house for the medical staf in which to live; and it is a great treat after the hard work of travelling during last week's fighting.

We are very busy. Men are wounded all over the place and I have seen some miraculous recoveries. For instance, one man got shot through the body from back to front, the bullet going through his right kidney, liver and bowel. He got peritonitis and was very dicky for some days, but now he is doing quite well. We have many medical cases that would amaze you.

There is another move shortly to try and capture DE WET. Once he is taken, they say BOTHA will give in."

SOLDIERS OF THE FOURTH CONTINGENT
Back: Alexander Farrell — John R. Hoare
Front: Frederick J. Kemmis and Henry Poppleton

THE FOURTH VICTORIAN CONTINGENT

A visitor to St. George's Hall in Donald on the 20th. of March, in the year 1900, might have wondered why forty young men were jostling each other around the platform in front of a distinguished-looking gentleman. He was Mr. J. H. Dyer M.L.A. and he was there to examine these strong-looking countrymen who had volunteered to join the Fourth Imperial Contingent.

However, he only needed 32 candidates from the Donald district. Therefore he dismissed anyone who was married, or under 21, and then found he was left with exactly 32 men. As they were all the regulation height, Mr. Dyer issued each man with a railway pass and told all of them to proceed to Melbourne immediately where the final selection would take place.

Amongst these 32 volunteers for the South African War was FREDERICK JAMES KEMMIS, the younger brother of William Albert Kemmis. He was ten years younger than Albert, but actually volunteered to go and fight for his Queen and Empire before Albert did.

Other Donald men who joined at the same time were Alexander Farrell, John William Neyland, John Richard Hoare, Thomas Morris and James Powell. Each one passed the strict tests for this contingent — which after all was raised for the specific purpose of sending **"seasoned bushmen who were bold riders, sharpshooters and capable of successfully contending with a fierce guerrilla enemy"**. There are no personal writings from Corporal Frederick Kemmis, although he must have been a good soldier to gain promotion and he would have had many close encounters with the Boers. Maybe he was there when three men from his regiment were viciously shot down at Zuilfontein, on 21st. April, 1901: -

> *"Before entering a narrow pass (between high kopjes) Sergeant Hurst and two men, Goebel and Clay, were sent out to reconnoitre not knowing that about 100 Boers were holding a small ridge along the track. The Boers opened fire at close range. Hurst and Goebel were riddled with bullets and Trooper Clay was severely wounded. Hurst and Goebel were brutally killed. Their bodies had to be brought nine miles down the pass on horseback and they were buried under a blue gum tree."*
> **(Warraknabeal Herald)**

Corporal Frederick James Kemmis served his Empire with a noble purpose and a brave heart; but he was only one among many Australians who-

> *"displayed their prowess as soldiers in a manner that somewhat startled the world and the memory of their work in South Africa will live as long as the English language is written."* **(Colonel Otter)**

After sixteen months on active service in South Africa, Corporal Kemmis was met at Cope Cope Railway Station on the 10th August, 1901, by a large number of local residents. As his train drew up, the local Riflemen fired several volleys in salute and the corporal received an enthusiastic welcome as he alighted onto the platform. The riflemen escorted him all the way to his home at Avon Plains where, at the invitation of Frederick's father, the company partook of an excellent meal and everyone expressed great pleasure to see Frederick at home.

Later, the Avon Plains riflemen organised a social in the Banyena Mechanics Institute to further welcome their old comrade back. Patriotic songs were sung and Corporal Kemmis was presented with a Lee Enfield match rifle.

In September, 1907, the Avon Plains Rifle Club elected William Albert Kemmis as their Secretary and Treasurer in place of Mr. F. J. Kemmis who was leaving to go to the Temora district of New South Wales where he had purchased land. Sadly, Frederick Kemmis was only farming in Temora for ten years when he died suddenly and tragically.

KEMMIS BROTHERS DROWNED

A sad drowning fatality occurred in the River Yarra on Sunday afternoon, the 30th. of September, 1917. Two brothers, George Kemmis, of Coburg, and Frederick J. Kemmis, of Temora, accompanied by a lad, obtained a boat to row across the river in order to visit the Botannical Gardens. When near the opposite bank the boat began to fill with water and the occupants, in shifting their positions, oveturned it. The lad swam safely to the bank, but the two brothers immediately sank and although efforts were made to rescue them these were unsuccessful. The bodies were subsequently recovered. Mr. W. A. Kemmis, of Avon Plains, is a brother of both the deceased.

(Donald Times — 2/10/1911)

CHASING JOHNNY BOER

The following letters written by L/Cpl. JOHN RICHARD HOARE, of Mount Jeffcott, describe his personal experiences of the Boer War. It is important to realise that big battles at the beginning of the war took place in main arenas, but after one year the war developed a very different aspect as it faded away into a seemingly never-ending pattern of continuous guerrilla attacks. At the same time it also faded away from newspaper headlines, although our brave soldiers were still losing their lives in the many bitter, small conflicts.

"It is **Christmas Day, 1900**, and we have been having a rough time lately with both the weather and the war. We are constantly on the march. The wet weather has set in over here in earnest and we are quite used to being in wet clothes for days on end. Even at night we have no shelter over our heads from the ceaseless rain.

We had a hard march last week when escorting a convoy of waggons from Lichtenberg to Otter's Hoop. We were 70 hours on the march with only six hours' rest and raining all the time. We went out again yesterday to escort a convoy to Zeerust (about 50 miles north of Mafeking) and suddenly we met about 300 Boers so we had a bit of a fight. There were fourteen Boers killed and wounded but we did not lose a single man.

We had a sharp fight at Driefontein on the 19th. December. Our column of 350 men left Lichtenberg at midnight and made a flank march around the Lichtenberg Commandos, while another column of our men came up behind their laager thus driving about 700 Boers towards us.

The enemy fought bravely, but just as the battle was turning in our favour some of the English gunners fired upon us by mistake with their 15-pounders, so we had to retire and we were very lucky no one was killed, though there were some very narrow escapes.

We will be shifting further down country soon and we will not be sorry. This is a very unhealthy part of South Africa where blazing, drought-stricken summers follow freezing winters. A great many of our men are very bad with enteric dysentery and several New South Welshmen have died from it here. I have kept in good health so far and have had many exciting experiences.

I have just taken my best mate to hospital. He has enteric fever and so I have taken his place as Orderly to Lieutenant Parkin, so from now on I will have no time to get into mischief."

In scouting duties the Australians proved to be cautious, reliable, self-confident and able to beat the Boers on their home ground and at their own game.

"It is now January 3rd. 1901, and we have reached that famous place called Mafeking. The weather is still very wet, but we are quite used to lying in the rain for hours while out scouting. When camped at Mafeking we can go down town where we do justice to a hot, comfortable breakfast after being out in the bitterly-cold nights.

I am kept busy as I am still Orderly to Lieutenant Parkin. He is a grand fellow to be with and I would follow him anywhere. He often takes my rifle to give me a spell. The last time we were out against the Boers, I fired 100 rounds at them from one position and then 60 rounds from another.

We are to leave this district tomorrow and will go down to Vryburg (about 100 miles south of Mafeking). The Boers are very active about there so we are taking a very strong force. You will no doubt see in the newspapers some accounts of what we are doing. Lord Methuen asked especially for 200 Australians to join him, although he could have had 500 English Yeomen if he wanted them. He prefers us and that makes Tommy Atkins jealous of Australians.

The English Gunners and Infantry prefer we Australians to do the Scouting ahead for them rather than the Paget Horse or the Yeomanry, as they reckon we are the only ones who understand this game; and a very exciting game it is, too, as we never know when we may be fired upon by the hidden Boers from behind huge rocks on top of the kopjes.

We are not allowed to shoot back when on Scouting duty because our task is to find out where the Boers are camped and then report back this information to our leaders. Then we all march through the night and surprise the enemy in their camp at daybreak. We then capture their ammunition and food supplies as they do not seem to want to fight when we attack their camps, but run away."

PROFILE – L/CPL. JOHN RICHARD HOARE

Hoare was born on the 15th. February, 1877, at Mount Jeffcott, near Donald. As a young man he worked hard on his father's selection until in March, 1900, at the age of 23, he volunteered to fight for his Queen and Empire in South Africa. He gave his address as Wood Street, Donald, and his religion as Roman Catholic. He was over 5ft. 7ins. in height. His interesting letters from the war zone reveal an observant, keen mind and a literary ability to describe events with accuracy. L/Cpl. JOHN RICHARD HOARE was discharged from the army on 25th. of August, 1901, after fighting in South Africa for more than a year and returned home to work on his parents' farm, near Donald. In 1908 he married Agnes Kate Doran and purchased a property at Sea Lake, but later moved to a farm at Tiega. J. R. HOARE and his wife had ten children.

> "We came into Otter's Hoop yesterday from Lichtenburg and arrived without any encounter with the Boers. Two nights ago we were told there were 700 Boers back in Wonderfontein, so we had orders to get up early and go to see about it. When we arrived there the "birds" had flown. Of course, we were disappointed, but we will soon be at them again and I shall give them a severe bump or two before long. Recently we had a good opportunity, but failed again. It happened this way.
>
> The Boers in the Lichtenburg district wanted to commandeer a young fellow to join their force and fight for them, but his mother had other ideas and would not

let her son go. Instead she sent him up to our camp to inform us of the secret location of the Boers' laager.

The lad came to our outpost at 8-30 one night and I was on duty so I halted him. He insisted that he was our friend and in broken English asked me to take him to someone in authority who could speak Dutch as he had some news of great importance.

So I took him up to the General for examination and he told us exactly where the Boers had their camp. Orders then came for every available mounted man to be ready at midnight and we were all marched out. At daybreak we were alerted by the sound of three shots coming from the enemy's outpost so we galloped in that direction.

We climbed a hill and from the top we could see about 200 Boers saddling up their horses. Obviously their sentries had spotted us and warned them. Immediately the two guns we had brought with us opened fire on them and our shells fell right in amongst them.

But before our gunners could fire any more shots their guns choked and would not fire another shell. Then we were told to gallop after them, but it was too late. The Boers were already on their horses and galloping away. If only they had let us attack sooner we would have captured most of them, but the officers kept us back behind the guns until it was too late.

Lord Methuen made a speech this morning at Church Parade and gave us great praise for the work we had done.

PRAISE FOR THE COLONIAL BOYS

When Lord Methuen first saw us in action he said, "Why did I not have such good men long ago? They are better than any guns. They will drive the Boers out of their cover." By this he did not mean Victorians alone, but men from **all** our Australian colonies.

He seems to take a great interest in us and he will come through the lines and ask how we are getting on for food and if we are satisfied. When it was raining at Lichtenberg he came around to see how we were getting on. The Colonel told him we were all right.

Lord Methuen replied, "Well, Colonel Kelly-Kenny, you may be all right in your dry tent, but these men are certainly not all right. I intend to look after my men so you turn out and come with me and we will find shelter for them." He did, too! We were all put up at a big store and were very comfortable until we left. Lichtenberg is as nice a little town as any we have struck in South Africa yet."

The Boers used "Hit and Run" tactics which were proving very effective, especially as the British soldiers travelled in long, slow-moving columns. It was very tiresome and frustrating for them, chasing an invisible enemy over and over again across the same piece of ground. British Infantrymen may have been highly-trained with modern rifles but what use were guns against an invisible foe?

L/Cpl. HOARE, of the Fourth Bushmen's Contingent, continues to play a brave part in a war that seemed to drag on indefinitely across that hostile terrain known as the VELDT. His letters are cheery in spite of danger and discomforts.

> "Our convoy of waggons and troops was over three miles long. General Carrington had twice failed to get the convoy through, so we expected a lot of trouble from Johnny Boer. We started off at 8-0 p.m. and travelled all night and did not pull up for a rest until 10 o'clock the next morning.
>
> Some of us were sent out as scouts to a kopje about two miles ahead to see if it was clear of Boers and we were very surprised to find bullets landing at our horses' feet. They fired a good many shots at us but no one was hurt. We could not see where the bullets were coming from — and that is the worst of it over here.
>
> As we rejoined our convoy, the Boers seemed to open fire on us from all sides, but there was no damage done. Unfortunately, we only got a few shots at them as they retreated. We finished our march after being 24 hours in the saddle without a rest, so you may guess we were all pretty sore and tired out; and to make matters worse it came on a terrible storm and every stitch we had on got wet through, so we had to make a big fire and sit around it until daybreak.
>
> We have had no tents for a long time now and sleep in our wet uniforms. We have had a rough time of it lately as there have been terrible hailstorms — and the thunder

and lightning are dreadful.

Last Saturday I met Will McPherson who used to be a butcher at Warracknabeal, and we were both pleased to meet each other. He is just recovering from Enteric Fever, but has grown into a big fellow, over 13 stone. He had some very narrow escapes from shells while in the trenches at the siege of Eland's River where many men died from typhoid fever which they caught during the siege.

Our Bushmen have gone to Mafeking to be refitted with new clothes which they badly needed. You might have seen in the newpapers that Lieutenant Gilpin was shot, also Woodman, on the 20th. August (1900). For a few days we stayed guard over the kopje where they were buried together. You need not be uneasy about me as I never felt better in my life and am getting fatter every day."

L/Cpl Hoare is serving in Lord Methuen's force and moving about incessantly through the broken countryside that lies between Rustenburg, Lichtenburg and Zeerust, chasing strong Boer commandos, winning small skirmishes, and suffering the indignity of the enemy's continual sniping.

"Our last four days were 'Forced Marches' as we were badly needed at the front and there was no time for delay. We stayed a day at Bulawayo and then came on by train as far as Crocodile Pools where the Boers were expected to come for water, but they did not come so we left a lot of artillery there to welcome them if they appeared and we came on to Mafeking. We are going out tomorrow to meet a force of the enemy. They are doing a lot of sniping around here; two of our sentries were killed last night.

We will have a day of reckoning with the Boers for this nasty sniping business. They shift in the night from kopje to kopje and one never knows where a bullet is coming from. We have had to grin and bear it, but later on we hope to pay them back for their nasty tricks — and with interest.

However, one company of our colonials had two fights recently and came out on top. Colonel Plumer says we are the most dare-devil set of men he ever saw; and we all think he deserves a lot of credit for the way he fought around here before Mafeking was relieved.

You probably heard how Colonel Plumer and his small Rhodesian Regiment (less than 700 and with no proper guns) bravely tried to cut their way through a large army of Boers to relieve Mafeking, but suffered heavy losses in their attempt, yet courageously managed to get within five miles of Mafeking before being driven back."

Reunion of Boer War Veterans

John Richard Hoare ~ second left, front row

BACK FROM THE WAR
— HOARE, MORRIS AND POPPLETON

On 26th. July, 1901, the local Donald newspaper reported that three more of our soldiers were safely back from the war. They were Lance-Corporal Hoare, Lance-Corporal Morris and Private Poppleton who arrived at the Donald Railway Station on Tuesday afternoon where a very large crowd of people assembled for the purpose of welcoming the three men back from South Africa. An impressive detachment of Victorian Rangers, under the command of Sergeant Hannah, and members of the Donald Fire Brigade, under Captain J. Crone, were drawn up at the railway station for the reception of the returned soldiers.

As the train steamed into the platform, the Donald Brass Band struck up with "Home Sweet Home" and a rousing cheer was given for the returning soldiers. After greeting their family and friends and being welcomed by the Committee of Reception, they were escorted to the drag and a procession formed behind the vehicle, headed by the band. Then everyone marched to the Post Office in Donald's main street where Councillor James R. Hornsby, President of the Donald Shire, welcomed the young men back to the municipality. Lieutenant-Colonel Basset, on behalf of the

Victorian Military Forces, welcomed the soldiers back to Donald and said if a call was made again to defend our shores a magnificent response from Donald would always be given.

In the evening a concert was given in St. George's Hall in honour of the young soldiers' return. President of the Shire, Mr. J. R. Hornsby, said that our Australian soldiers had earned a good name for themselves and their country. They had gone to South Africa to fight in the cause of liberty and justice and not for their own personal gain.

Then Lieutenant-Colonel Basset introduced the audience to the three returned soldiers — Lance-Corporal Hoare, of Donald; Lance-Corporal Morris, of Cope Cope; and Trooper Poppleton, of Donald. They had enlisted in the Fourth Imperial Contingent which landed back in Melbourne by S.S. ORIENT on 12th. July, 1901. He congratulated the men on their safe return and invited them to speak. First, Lance-Corporal Hoare briefly recapitulated the most memorable parts of his campaign in South Africa (see his letters for this account).

Trooper POPPLETON then gave his experiences, speaking especially of their severe fight at Haartebeestefontein. He was fighting under the hard-working Lord Methuen. The Australian Bushmen and English Yeomanry drove back De Villiers and his Boers from Bechuanaland and cleared the country as far as Vryburg. From Taungs they had crossed the Transvaal border and made for Klerksdorp. They successfully brushed the enemy aside at Uitval's Kop, Paardefontein and Lilliefontein; but on February 19th. they met a considerable enemy force under the command of De Villiers and Liebenberg at a place called Haartebeestefontein.

The day before they had pounced upon a Boer laager and captured 10,000 head of cattle, 43 waggons and 40 prisoners. Inspired by this success, Lord Methuen now attacked the pass being held by the Boers; but the enemy had a larger number of men and were entrenched in a very strong position.

Trooper Poppleton described how it took them five hours of hard fighting to dislodge the Boers from this narrow passage through the mountains; and although Lord Methuen only had 1,500 men against a considerably stronger enemy, they finally won the fight. The Yeomanry, the Australians and the North Lancashires all did well and were highly-praised by Lord Methuen. British casualties amounted to 16 killed and 34 wounded, while the Boers left 18 of their dead upon the mountain slopes. Trooper Poppleton added these words: -

> *"This was where they said Poppleton was a cocktail (coward). I'll tell you all about it. The advance scouts were ordered into the firing-line. I belonged to the scouts, but at that time was on duty as an orderly to one of the officers and could not get away at once. I followed them up as soon as possible and had almost reached the firing-line when an officer galloped up and ordered me to take a message to another officer who was in charge of the flankers. Just as I turned my horse to do his bidding, a 15-pounder shell came over us and a few of those nearby said, "Hello, Pop's doing a break!" They thought I was a coward and running away. However, I safely delivered the orders I had received and then returned to the firing-line just in time to see one of the finest fellows in our*

Fourth Contingent shot down and killed. I was so mad that I fired all my bullets in the direction I thought his assailant was. I advanced as close as possible in the hopes of hitting his killer."

Lance-Corporal Morris then recounted some of his wartime experiences. He said he was not in the same part of South Africa as the other two, but had seen a good deal of fighting for all that. His first experience of being under fire happened when he was out on patrol, riding through a narrow pass with thirteen of his mates, and the enemy opened fire on them.

"We did get off our horses a bit quick and I gave my horse a dig with my rifle to move him away and then got into a gutter and started firing back. We could see rifle flashes at the top of the hill so aimed at that spot. Luckily, our supports came up after a while and we were relieved. Next morning we climbed up and looked at the place we had been aiming at since no one else had been shooting at that spot. We found five dead Boers. Who shot them if it wasn't us?

Another time I was out scouting in a party of twenty when we discovered a body of 1500 men camped in a valley, but we could not tell whether they were Boers or British as the Boers often wear captured khaki uniforms.

Two of us volunteered to go nearer and find out, so we rode down the kopje and gradually drew nearer to the camp until we saw two sentries. Then we heard Dutch being spoken so I rode nearer to them and tried to persuade one of the burghers to come over to me, but he would not.

Then I knew I was in trouble, for if I turned and galloped away they would fire on us for certain, but if I walked they might be deceived. So I turned my horse and moved away quietly. I had not gone fifty yards when I heard the sharp crack of a Mauser rifle and a bullet whistled over my shoulder.

Then I really set sail! For at least 400 yards we were perfect targets for the Boers. They were rotten shots for they only succeeded in putting two holes in my overcoat as it flapped around me."

PRIVATE THOMAS MORRIS

Thomas Morris was a young bushman, living at Cope Cope and working for his brother-in-law before he joined the Fourth Imperial Regiment. Several other Donald and district men joined the same contingent which left Melbourne on the 1st. May, 1900, on the transport "VICTORIAN". They disembarked at Biera about three weeks later on the 23rd. May.

His enlistment papers show that Private Thomas Morris (No. 518) was born on the 26th. October, 1876, and was six feet tall. By the time he returned from the war he had been promoted to the rank of Lance-Corporal.

He wrote to his sister from Bulawayo where his regiment had been protecting the lines of communication: -

> "I am letting you know that I am still alive. When we first arrived here we had to do a long march from Marandellas to Bulawayo; it was 290 miles and it took us 22 days to do it. We could have done it faster if they would only feed our horses, but they only give them about two handfuls of oats and mealies, three times a day, so the poor things were nearly dead when we got there. My regiment is scattered all over the place acting as escort to the different columns.

Two of our men got shot the other day, an officer and a trooper, but I don't think there will be many more casualties as the war is about all over. We had a rifle match the other day and my group won easily and got a prize of five pounds. I am not going to stay here when the war is over. You would want a big fortune to buy land and cattle here, and there are so many diseases among the stock that they die off in great numbers every year. The grasshoppers are as thick as bees and as large as birds.

There are plenty of lions around Marandellas but they are not very savage and do not come near the camps.

This is a cold place in winter, but very healthy; about the only thing a chap has to be frightened of is dysentery and it has been severe on some of our boys. I have had a touch of Malaria. It's a nasty thing and goes through all one's joints and mostly to the head, nearly driving you crazy. Our horses were dying of glanders; they are full of ticks and lice.

On our return to camp the other day our convoy was about 6 miles long and there were two of our flankers killed by Boer snipers and two more had their horses shot underneath them. We burnt all the Boers out along our way, turned women and children out of their houses, gave them ten minutes to get their goods and chattels out, and then burnt their houses to the ground.

I suppose you think this is a bit rough, but it is good enough for them as these houses belong to men who handed in their weapons and swore allegiance to the Queen, but now they are traitors and fighting against us again so it is the only way to fix them. You ought to see the "Tommies", they are pretty good at looting fowls, pigs, and sheep as the poor chaps need the extra rations.

The Boer women are nearly as bad as the men; they will not attack a large party, but will shoot down our patrols like dogs. In scores of places our men have been shot down by bullets from the windows of these very same houses.

Excuse the state of this letter as I am writing it on a rock, and in the dirtiest place under the sun. In fact, our eyes are nearly cut out with the dust; and some men get the dust in their lungs.

Around here the good patches of land are in the gullies. Zeerust was the best place for magnificent oranges, lemons and limes. I commandeered them, filling up my haversack two or three times with the juicy fruit."

TROOPER CHARLES MIDGLEY OF MINYIP

Guerrilla warfare in the Transvaal continued during the rest of 1900 and throughout 1901. De Wet continued his sporadic, successful attacks upon the British convoys and railway trains, keeping our men very busy trying to catch him. Moving about through the Transvaal were several other strong, mobile Boer commandos led by Commandants Lemmer, Snyman and Delarey; and the British columns chasing them were led by Lord Methuen, General Douglas, General Broadwood and Lord Enroll. However, Lord Methuen's force contained a large proportion of our Australian bushmen and amongst them were the two Midgley brothers from Minyip. This following letter from Trooper Charles Midgley to his sister gives the reader a clear picture of guerrilla warfare with all its horror: -

"**Lichtenburq, November 13th, 1900**. — *A few lines to let you know our whereabouts and what we are doing. We have had two fights recently with Lemmer's commando, giving him a belting each time. Last month we caught him unawares with an early dawn attack. We had to travel by night, leaving Kafir Kraal at 1 a.m. for*

Wonderfontein where we knew a Boer laager had been sighted.

We came upon them at daylight, just as they were preparing breakfast. Then the fun commenced and lasted for three hours. We chased them in all directions; eight were killed and forty taken prisoners, besides which several waggons, mules, Mauser rifles and souvenirs were secured. One of their killed was Lemment, a field-cornet from Buffel's Hoek. Eight of us, including our colonel and lieutenant, got separated from the main crowd and had to do any amount of riding to escape the bullets flying all around us.

We got out of it alright with only one man being wounded in the arm. Three horses were wounded, one of them got a bullet through its head just above the eye, but he is still going strong. They used a lot of explosive bullets which sound just like the crack of a stockwhip when they burst.

Well, we left the camp at 3 p.m. for Lichtenburg and suddenly came upon the Boers again about two miles from the town. They were waiting for us and had their big guns blazing away, but they must have been flurried as not one of their shells did any damage; and then they ran away for their lives when our artillery got going. We captured a pom-pom from them and now they are hiding in the kopjes as usual, all over the place.

They were parading around the streets of Wonderfontein before we arrived, so we soon gave them a change of climate. This town is well-named as it really is like a "wonderful fountain" situated on a plain, in a crater, and the water is as clear as crystal.

Poor beggars, I sometimes feel sorry for them as they never know when we are going to disturb their meal. At Wonderfontein, their boots, socks, saddles etc. were strewn about, they not having had time to dress, and their kettles were boiling at a great pace when we took possession of their camp. They are an ignorant crowd letting themselves be urged on by a few strong men like Botha, Lemmer, Delarey and others. But their leaders know perfectly well that if caught they have only two choices, either killed or transported.

We are now eating off their crops and animals which we have commandeered. You may depend on it that we like getting into the towns. Why, I actually had roast pork, mashed potatoes, onions and green peas for dinner yesterday and we are having fowl or pork every meal now. However, the food does not last long divided up amongst a crowd of two or three thousand soldiers. Lord Methuen always looks after his men; so as we are without tents he found us a place in town and we are now sleeping comfortably in a deserted store and using crockery."

This contemporary cartoon pays tribute to a tenacious soldier, General Charles Knox, who was determined to frustrate De Wet's plan to invade Cape Colony. At the beginning of December (1900) De Wet with 3,000 men and several big guns trekked southwards with the intention of crossing the Orange River into Cape Colony where he would stir up trouble against the British. However, General Knox went after him in full cry. His men rode through dreadful weather, their horses struggling in deep mud, heavy rain lashing their faces, yet they reached the Orange River in time to stop De Wet and make his army turn back to the north.

Trooper Alexander Farrell

AUSTRALIANS AND THE NORTH LANCASHIRES

Private ALEXANDER FARRELL, of Laen, became a mounted infantryman serving under Lord Methuen. In a letter to his family he explains how, after crossing the Transvaal border on their way to Klerksdorp, they were constantly fighting the Boers.

Lord Methuen's force consisted largely of Bushmen and Yeomanry because that hard-working general preferred using colonial mounted troops for his purpose of clearing the area of roaming Boer commandos. Private ALEXANDER FARRELL took part in all the battles and skirmishes that occurred as they worked through a difficult area known as the Masakani Hills, but they successfully brushed the enemy out of their path.

Then on February 19th. they reached Haartebeestefontein and came upon an army of Boers who had taken up strong positions for battle in order to prevent Lord Methuen's progress. After many hours of hard fighting the Boers were successfully beaten off from the pass which they were holding and Methuen's force victoriously reached its destination.

6/3/1901. *"We have had a real good time of it with "Jacky the Boer". After we left Lilliefontein we were fighting all the way till we got to Klerksdorp. Our best fight was at a place called Haartebeestefontein and General Methuen said it was the hardest fight he has had since Magersfontein. It lasted six and a half hours and all the time we poured a continuous hail of lead and shell into the Boers.*

It was the first time I have been under shell-fire from the enemy and I can assure you I never want to get under another. It's no wonder the Boers clear out as soon as our guns begin to shell them; and fortunately for us, the shells from their 9-pounder Krupp guns sailed harmlessly over our heads.

Out of the six shells that the Boers fired at us, only one exploded, killing four sheep alongside one of our waggons. The rest of their shells were all plugged up at the nose with wood; which just shows that the Boers are short of ammunition because they are just re-filling old shell cases.

Their shells cannot do us any harm unless they strike direct, as the Boers are not expert enough to fuse the shells to enable them to explode. We (the Victorians) have lost five men (two troopers killed, three troopers wounded and three officers wounded). Lt. Colonel Kelly, Lt. Parkin and Lt. Mann are dead.

Sergeant Vaughan was shot dead only fifteen yards from the Boers, so that proves how close we got to them. All our men who were killed and wounded were shot with explosive bullets — and they make a terrible wound.

Well, we were so close to them that until the North

Lancashires were brought up on our left we dared not move once we got on top of the kopje. To see the North Lancashires advancing up that kopje without the least bit of cover, and never flinch, was a sight I shall never forget as long as I live. They were under heavy fire all the time till they started to climb the kopje.

Then the Boers had to come out of their hiding places behind the rocks in order to see the advancing Lancashires: and we being on top of the kopje to their right, caught them every time they rose, mowing them down like grass, until at last they bolted.

All this time our dear old general was lying down alongside of us; and as soon as the Boers started to retreat General Methuen clapped several of us on the back saying, "Well done, my lads, you have fought most nobly."

General Methuen may have had a severe set-back at Modder River and been abused for it, but ask any of his men what they think of him and see what answer you will get. Colonials swear by him!

As for myself, I would not ask to serve under a truer, nobler, or better general. He is every inch a soldier. Would he send his men on to a kopje where he would not go himself? No! Where his men are, he goes, too, no matter how hot the fire.

I have since been in hospital at Johannesburg for ten days with a slight attack of fever and rheumatism in my right ankle. I am alright again now and off to rejoin my regiment this week.

The Boer losses in the Haartebeestefontein fight were estimated between 150 to 200, whilst our losses were

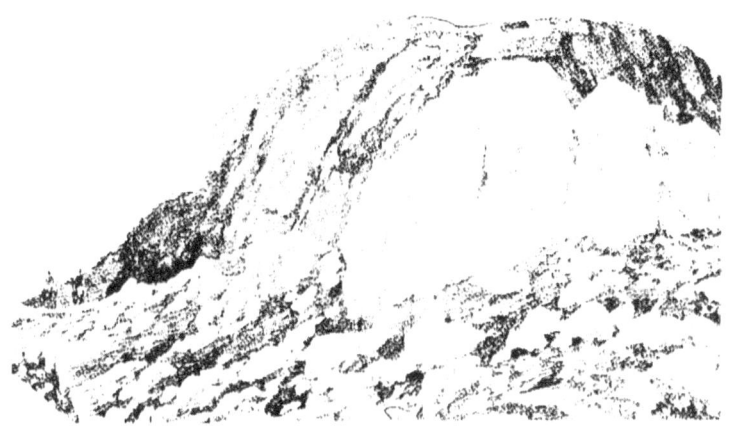

They were fighting in a country dotted with rocky kopjes (hills) which had the most grotesque shapes.

sixteen killed and thirty-four wounded. The Boers are desperate for supplies so whenever they capture our men they strip them of their uniforms and accoutrements before releasing them. Then our poor fellows are sent back to camp, naked and blushing, with the compliments of De Wet.

Recently Lord Kitchener issued strict orders that any Boer caught wearing a British uniform is to be shot immediately."

Lord Kitchener's anger was quite justified because on several occasions British soldiers were deceived by the Boers' ruse of wearing stolen khaki uniforms, a trick which enabled the enemy to approach within yards of a British troop and so gain a crucial advantage in any ensuing battle. By wearing stolen uniforms the Boers were able to ride up to close a British camp and shoot down a number of unsuspecting men before any preparations for resistance could be made. This nasty trick never seemed to fail and the Boers often made successful use of it. However, the Boers went too far one day when they were seen wearing Scottish uniforms for the Scots could never forgive such an insult and swore terrible vengeance upon the Boers for daring to desecrate their sacred kilt in this way.

The uniforms and slouch hats worn by both Boers and Australians were so similar that quite often in the mist or darkness it was not easy for Tommy Atkins to distinguish between his friend or foe, hence the added danger to our Colonials of being shot down by their own side.

PROFILE – ALEXANDER FARRELL

Farrell was born at Laen North on the 3rd. October, 1875, the son of an early settler named Alexander Farrell. He grew up on his father's selection and they worked hard to make a success of farming in spite of droughts and rabbits. He knew how to ride and shoot — essential skills for a bushman. When he volunteered to fight in South Africa he was just 25 years old and his enlistment papers describe him as "single, a farmer, Anglican, nearly 5 feet-9 inches tall with a chest measurement of 36 inches". No. 526 Private Alexander Farrell left with the Fourth Imperial Contingent on 1st. May, 1900, in the transport "VICTORIAN". After completing his twelve months' service, he was invalided home to Australia, travelling on the transport "ORIENT". He felt so ill when he arrived in Melbourne that he went straight to the Melbourne Coffee Palace to rest. In his room he collapsed on the floor in an unconscious state and would have died if his friend had not found him in time. L/Cpl John Hoare saved Alexander's life by calling a doctor and then taking him to the Melbourne Hospital. Alexander stayed in hospital for several months and was not officially welcomed back to Donald until May, 1902, after a serious and protracted illness. Eventually he was granted an army pension. He returned to work on his parents' farm. On the 15th. May, 1906, Alexander Farrell married Sarah Cathcart, a neighbour's daughter. They had two children, Alexander (Lex) and Gladys; and in 1926 Alexander purchased land at Watchupga and took his family to live in the Mallee.

A Typical Scene From The War Alexander Farrell and his friends are photographed by a supply train and behind is a kopje from which Boers are probably watching them. Notice how the uniforms of our Bushmen make them look like the Boers, especially the slouch hats.

PRIVATE H. SCHOMAN

"The Wilmansrust Disaster" by "One Who Was There"

HERMAN SCHOMAN was born at Laen and he was just one of several young men who left the farming district of Donald to go and fight in South Africa. He enlisted with the Fifth Mounted Rifles Contingent which left Melbourne on 15th. February, 1901. In a letter to a friend in Donald, Herman tells a sad story of the disaster which overtook his section of the 5th. Contingent on June 12th. 1901, at a place called Wilmansrust — situated between Middelburg and Bethel. At the time, Private Schoman's detachment was under the command of Major Morris.

> "**Middelburg. 1st. July, 1901.** *Dear Dave, We are having a fairly good time of it since we have been doing a bit of scrapping with the Boers now and again. We were getting on very well until about two weeks ago when we got a smash up! We had been skirmishing with the enemy for three days and then we camped one night.*
>
> **We had two Boer guides with us, but they were traitors and went out and led Viljoen's party into our camp.** *The Boers got within about 30 yards when they poured in two volleys; and then with a yell like a lot of blacks, they rushed the camp.*
>
> *Most of us had just turned in and our rifles were stacked so that we did not have a chance. They rushed in amongst the lines of sleeping men, firing at anything they*

saw moving, and singing out, "HANDS UP KHAKEE!"

A lot of chaps got shot with their hands up, and a lot got shot in their beds. It was murder all right and I don't know how I got out of it alive. Four of my best chums were shot dead. There were about 160 of our horses shot during the attack and it was the horses that saved us as we sheltered behind them.

We had twenty-one Victorians killed that night (one has died since) and forty wounded. We do not know how many of the Boers were lost. We went to look but they fired at us, and that was enough. It seems like a dream, it was so sudden. I won't foget it in a hurry.

I was in a warm corner once before, but it was not a patch on this latest massacre. That time, 500 Boers fired into our camp at 1400 yards. They did rock it for fifteen minutes, but they did no damage, only killed a nigger and wounded one of our men. We pegged six of them.

We have had very few battles, but we meet with plenty of snipers hiding behind their rocks on the kopjes. They are the worst danger. You can hear "ping-pong" — that is the sound of a Mauser — you can hear it, but it is hard to find the place where it comes from. We have had a few men wounded by snipers, and if we can place the snipers we give them "rats".

It is not a bad game, only for the long time they keep us in the saddle; sometimes from 3 a.m. till 12 the next day. It is long hours and it is very cold now that winter is here."

Lieutenant Frederick STEBBINS was one of the officers with this unfortunate party of the 5th. V.M.R. and he told the same story of treachery and mayhem. (It seems the British were certainly learning the importance of secrecy the hard way.) Two other Donald men, William Albert Kemmis and Robert Alfred Watson, were also there and poor ROBERT ALFRED WATSON was one of the unlucky ones.

When the Boers suddenly charged amongst our sleeping soldiers, Trooper R. A. Watson may have been asleep, rolled up in his blanket, and therefore not quick enough to move away. He was one of those shot in their beds. Badly wounded, he was taken to Middelburg Military Hospital where he received medical care and attention. However, his constitution having been severely weakened by his wounds, he caught the dreaded enteric fever.

No. 1590 Private ROBERT ALFRED WATSON who had fought so bravely while serving with the 5th. Victorian Contingent, died at Middelburg on 6th. December, 1901.

Sergeant Niven Neyland
A soldier-hero of the great boer war and the first world war
Photo — Grace Neyland-Keating collection

PROFILE: PRIVATE NIVEN NEYLAND (NO. 1133)

Neyland was born in 1878, the second of thirteen children; and enlisted with the Fifth Victorian Mounted Rifles on the 9th. February, 1901, at the age of 22 years 6 months. He served for one year and forty-seven days with his regiment in South Africa fighting against the Boers, but apparently survived unscathed. His enlistment papers describe him as 5ft.-10ins. tall, with a dark complexion, brown eyes and black hair. When he left the regiment on the 28th. March, 1902, he expressed a desire to stay in South Africa and the following testimonial explains what he did there: -

> SOUTH AFRICAN CONSTABULARY Remount Depot. Hill Crest, Natal.
>
> Mr. Niven Neyland has been a Civilian Conductor at this Depot since the beginning of April last. I have great pleasure in testifying to his conduct, sobriety and capabilities and he has been of great service to me. He has been in sole charge of a continually changing stable of some 120 Australian Remounts and is a first rate horse-master and rough-rider and always willing to turn his hand to any job. **He and his two brothers** are leaving me on the abolition of Civilian Conductors in the South African Constabulary. I hope he may be successful in obtaining a billet in the Repatriation or other Government Services. He states he has served during the war with the 5th. Victorian Mounted Rifles.
>
> Captain W. T. Wilcox. (Nov. 1902)

Niven married Miss Linda May Pryse in 1906. They had four children, but then Niven volunteered once again to fight for his country in 1914. He was 36 years old and a farmer when he rushed to enlist on the 31st. Sepember, and he sailed to Egypt with the 14th. (Victorian) Battalion on 17th. June, 1915. He fought the Turks on the Gallipoli Peninsula and during the British offensive in August, 1915, he was taken prisoner, but only after a fierce resistance by his small detachment. Niven was wounded and his health ruined by the cruelty of the Turks while he was at their mercy. In November, 1918, he was repatriated and sent home. His fifth child, Joy Pringle Neyland, was born in 1920.

PROFILE: JAMES LESLIE NEYLAND

Neyland was born at Corack in 1880. He was the third Neyland brother to go to the Great Boer War (1899-1902). James Leslie   left Australia with the Third (Bushmens) Contingent on the "EURYALUS" as stated in the Donald newspaper on 9th. March, 1900: -

"The Bushmen's Corps will sail for South Africa on Saturday and amongst the men leaving with them are three from this district — Mr. James Neyland, of Corack; Mr. James Meyer, of Donald; and Mr. Roderick McSwain, of Boolite. (The last named gentleman is the tallest in the regiment.) It is to the credit of the Donald district that these men have been selected and the settlers up this way know their sons will be amongst the first to rush into danger; for when bravery and courage are wanted, the Donald boys will all be there."

James Leslie was presented with a fine pipe as a farewell

present from the Donald Community, together with their best wishes and hopes for his safe return; and his father gave him a fine horse to take with him to South Africa. James Leslie never returned because at the end of the war he stayed in South Africa with his brothers, John William and Niven, working in the horse department to restock the farms that had been depleted during that war. When his two brothers returned to Australia, James Leslie remained there as he obviously liked his job and the country. However, he often wrote home and sent viewcards and presents of ostrich feathers (popular for decorating hats in those days). At the outbreak of the First World War he was employed as an engine-driver with the Podgers' Gold Mining Company on the Rand, but his fine spirit once again persuaded him to do his duty for his country. So he enlisted under the command of General Botha (Britain's enemy in the Boer war) to quell the German uprising in South Africa. After that fight was over, James Leslie sailed for England in the hopes of joining other Australians en route to France. He stayed in Aldershot Army Barracks where his experience was much appreciated in the training of new troops and he was promoted to **Sergeant James Leslie Neyland**. However, he insisted on being sent to the front and was killed while fighting with the Gordan Highlanders in the Battle of Arras on 9th. April, 1917.

TRUMPETER CHARLES PEARSON

Charles Pearson wrote an interesting letter to his Donald relatives from a hospital in Fourteen Streams, a place on the Vaal River. It tells the circumstances of his mishap, but at the same time gives the reader a lively account of the Bushmen's attack on De Wet's convoy near Klerksdorp.

"**March, 1901. Fourteen Streams**. — *I an sorry to tell you that I am in the ambulance waggon with four holes through me; two on the right side, one in the left shoulder and one in the left leg. But the bullets passed right through me, thank God, and I am now allowed to sit up and read. I asked Sister Tiddy, of Stawell, to help me write this letter.*

I was wounded near Klerksdorp when we attacked De Wet's convoy and was shot while riding at full gallop over to the General with a message. I had to cross over a piece of open plain and the dust was flying up from the bullets striking the ground around me.

I reached the General and then carried back his message under rapid shrapnel fire, though this time I had some cover from the kopjes. When I tried to get off my horse I found I could not move and so I was lifted off.

The doctor dressed my wounds and I lay there waiting for the ambulance bearers. Then we heard rapid gun-fire which came from two of our armoured trains; and they gave the enemy "rats".

We had nearly run out of ammunition so it was fortunate that the trains came along and drove the enemy back into the hills causing them to leave their supply waggons behind. The shells from our trains were bursting in the midst of them, compelling them to retire so that they left their horses, dead men and wounded behind. This is one of the heaviest defeats De Wet has had during the war.

Then the Australian Bushmen, 700 strong, fixed bayonets and charged, firing as they advanced; but the Boers declined this more intimate acquaintance with cold steel so turned and fled to save their own skins.

Our 15-pounders got up closer and opened fire upon them. Then the Bushmen took up the chase again with disastrous results to the flying foe, many of them dropping to the ground. The Boers eventually disappeared into the hills and the Bushmen let them go; pursuit was abandoned because their tired horses could go no farther.

We captured all their rations and ammunition, 30 prisoners, 109 wounded, 91 killed and a string of their waggons. We escorted this lot to Vryburg where they were handed over to General Settle who had 3,000 mounted men with him and many of these men are colonials.

We now go to join General French who has Botha hemmed in a corner where we must make him surrender or he will lose all his men. We have only 300 miles to *go and we travel day and night, having about two or three hours' sleep now and again. We hope to put an end to this war somehow soon. All the railway traffic has been stopped, except for the armoured trains which are bullet-proof.*

You have no idea of conditions over here. Tremendous rains have fallen and the country is one huge quagmire. We have endured continual heavy rains, rough winds and rocky country, so that we are wet to the skin more often than not. Today it is burning hot, but we were in heavy rain when fighting De Wet."

Trumpeter CHARLES PEARSON was attached to No. 4 Battery of the Royal Artillery and bravely fought against the Boers. It was widely acknowledged that our Australian Mounted Infantry played a brilliant part in the expulsion of De Wet from Cape Colony. In fact, by the end of April, 1901, De Wet's invasion of Cape Colony had ended in disaster with all his commandos hunted back over the Orange River.

FIGHTING IN A GUERRILLA-STYLE WAR

The Boer commandos knew the lie of the land and they were able to collect food as they went along. They rode sturdy Cape ponies and could move quickly. Also, they could break up too easily to be overtaken and captured by the long slow-moving columns of British Infantrymen. All the time our poor infantrymen steadily marched their twenty miles a day, crossing and re-crossing the arid plains, winning their way at the expense of their boots and not their lives.

Most people thought that the British occupation of Pretoria (Kruger's capital)) on the 5th. of June, 1900, meant the end of the war, but they were wrong. Although General Botha and other Boer leaders were greatly in favour of peace, Christian De Wet and Delarey refused to surrender. They were both clever, audacious fighters who subsequently carried on an effective guerrilla campaign which dragged on this war at the cost of many more lives of brave men on both sides. No one thought the war would last another two years.

Thus began a new type of war known as GUERRILLA WARFARE. No longer were there great dramatic battles between two armies on an open plain; instead the Boers made scattered, small raids throughout the countryside to inflict misery in the hopes of vengeance upon the British and thus prolong the resistance. Boer commandos had eyes that never failed to notice the weak points of the British positions, especially along the railway lines where they were able to inflict cruel and expensive damage to lives and property.

However, the Boers avoided direct confrontation when possible. Their leader, Christiaan De Wet, told them, *"We have to use opportunities as they present themselves, at other times, flee!"*

An example of this opportunist strategy was De Wet's attack on the Rooiwal Railway Station (7th. June, 1900) when he burnt more than £250,000 of British supplies that had accumulated there. This kind of raid was very good for Boer morale, but very bad for the British taxpayers.

Farrier-Corporal Leslie PILKINGTON was on active service in South Africa when he wrote this letter describing army conditions and guerrilla tactics: -

Wonderfontein, Eastern Transvaal, 30th. April, 1901.
"It is nearly three months since we have been near the railway line, or slept in tents. We have been living for nearly four weeks on nothing but mealie porridge, and very little of that. We have no salt of sugar, and neither bread nor biscuits of any sort. It is a terribly rough life and bitterly cold at night with only one blanket to keep out frost and rain.

We have had a good deal of severe fighting, but so far I have escaped unhurt. We had a terrible time in February last, on the 11th. We were camped at a place called Lake

Chrissie with General Smith-Dorien's column, and the Boers attacked our camp at three o'clock in the morning (their favourite time for an assault). They shot down our sentries and attacked the sleeping camp. It was pitch dark and the confusion was something terrible.

To make matters worse, the Lancers' horses stampeded and galloped over us straight at the front line of advancing Boers. The enemy thought it was us charging them with the bayonet and stopped action for a minute. Well, that minute saved our lives as it gave us time to form up and give them a volley which killed seventeen of them. The Cameron Highlanders then made a splendid charge with bayonets which fairly routed the enemy and they ran away like frightened sheep.

It was a terrible sight. Men were lying, groaning and bleeding all around me. My greatest chum got a bullet through the neck and died a few minutes afterwards; he was right beside me all the time. Altogether we had 27 killed and 39 of our boys wounded.

We chased the Boers all the next day and captured a good number of their waggons and guns. I shall be glad when it is all over."

By means of a surprise night-attack, General Botha hoped to overwhelm a section of the British force that was blocking his road to the north. Through the dark and misty night, large numbers of the enemy advanced to make their deadly assault, but our soldiers bravely bore the brunt of that surprise attack and even succeeded in turning the tables. Corporal Pilkington's casualty figures show how heavy had been the Boer bullets which swept through his sleeping camp.

JAMES, LAURENCE AND RONALD MUIR

PROFILE – BOUNDARY RIDER
RONALD THOMAS MUIR

Muir and his three chums — Charles Routley, Percy King and John Hamilton — left Donald together for Langwarrin Training Camp on the 14th. January, 1901. They sailed to South Africa with the Fifth Contingent on 15th. February, 1901. No. 982, Private Ronald Thomas Muir, fought in many dangerous skirmishes while serving in that regiment and when the war was over he returned home to a hero's welcome.

> "**Ronald Muir** is O.K." wrote John Hamilton from Sunnyside Camp, Pretoria, on the 2nd. April, 1901.
>
> "**Ronald Muir** is in my Company and he is as fat as butter and resembles Andrew Muir as much as possible," wrote Percy King from Bronkhorst Spruit on 25th. July, 1901.

While he was in South Africa, Ronald Thomas Muir corresponded with a young woman named Ethel Hemley and soon after receiving his letters she became ill with a fever and a nasty rash on her face and body. Doctors could not identify her complaint at first, but later decided she was suffering from Smallpox. They were very puzzled about how she contracted the disease when there were no cases in Victoria, so asked if she had been receiving mail from foreign countries recently.

"Not really," she said," only a letter from a soldier fighting in the Boer War." Experts decided she must have caught Smallpox from Ronald Thomas Muir's letter and immediately a warning was issued by the Health Authorities with the result that in future all mail arriving into Australia from South Africa had to be fumigated before it was distributed. One can therefore thank Ronald Thomas Muir and Ethel Hemley for the new health precautions that were enforced to protect folks here. Ethel recovered from the terrible Smallpox, but a deep pitmark scarred her face for the rest of her life.

Ronald returned safely from the war in June, 1902, married Ethel Hemley and worked hard to support his family on a farm at East Laen. They had four boys and three girls; and when their youngest son was born, Ronald named him "Herbert Kitchener" in honour of the great British general. Sadly, Ronald died from influenza in 1923 leaving a widow and seven young children. His son, Herbert K. Muir, grew up to become a well-known landscape artist.

Ronald Thomas Muir

James Muir

MATESHIP
Back into the midst of shot and shell rides a mounted rifleman to rescue his mate whose horse has been shot from under him. He is one of many unknown heroes, risking his life to save a fellow soldier.

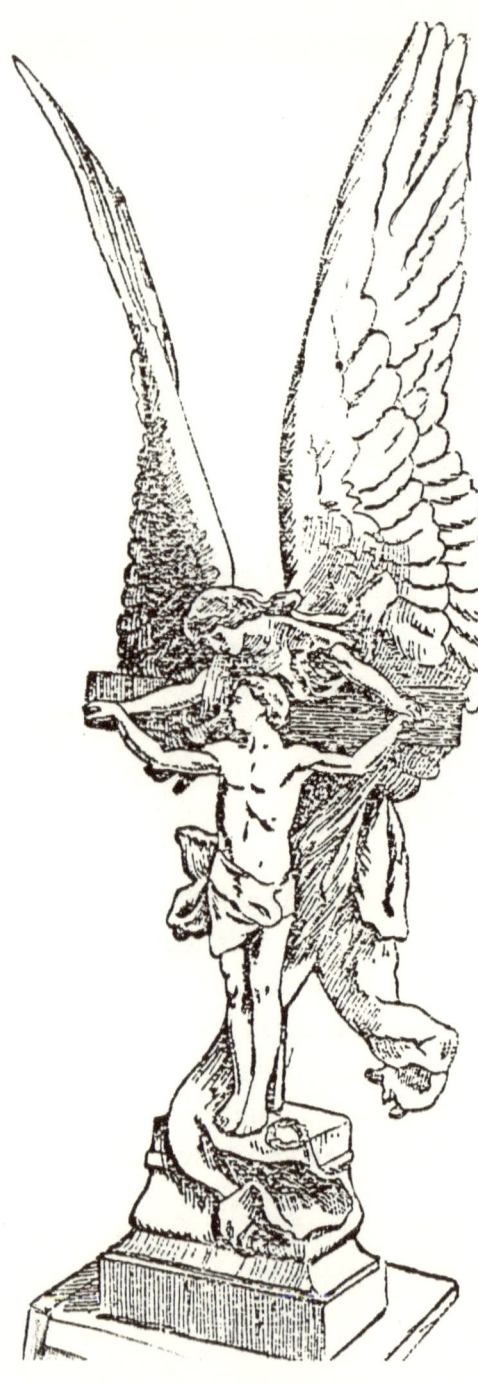

ST. PAULS CATHEDRAL — LONDON
Memorial to Colonial Troops who fell in the SOUTH AFRICAN WAR (1899-1902)

DONALD'S LAST HERO RETURNS HOME

On Friday afternoon, the 4th. of July, 1902, the last of Donald's volunteers for active service returned from South Africa. The town was colourfully-decorated with bunting as large numbers of people made their way to the Railway Station to welcome back CORPORAL LAURENCE MUIR when he stepped off the afternoon train from Melbourne. As the train drew to a halt, the local band struck up that emotional song, "Home Sweet home" and Corporal L. S. Muir was escorted to the drag. Surrounded by smiling relatives, he was driven to the local Post Office in a procession that was led by the band.

Councillor J. R. Hornsby, J. P. welcomed him back to Donald after his hard work on the field of battle and congratulated him on having worthily upheld the good name won by earlier Australian contingents.

At the 'WELCOME HOME" concert in the evening Councillor Hornsby welcomed back Corporal Laurence Muir and pointed out to the audience that the war was one of justice and equity, in which England was simply fighting for equal rights for all people and it was not undertaken for territorial aggrandisement, nor for gold as many who were opposed to the war would have us believe. Lieutenant

Laing, of the Victorian Rangers, welcomed home Corporal Muir and said the careers of our local Donald men had been watched with great interest and we were very proud of the splendid name earned for Australia by her sons.

After thanking everyone for their hearty welcome, Corporal Muir exhibited a number of articles which he had brought back from South Africa. He held up his LEE-METFORD rifle and explained its workings. Then he showed a Boer MAUSER rifle complete with bandolier. Next he showed the audience a number of ZULU weapons, including a beautiful little battle-axe. Finally, he held up a CAMP BLANKET and a WATERPROOF SHEET, two items which every soldier must carry on marches if he wants protection at night.

Laurence Muir volunteered again in 1914 to fight for the cause of Freedom and to save his country from German oppression. He was killed in action at Villers-Bretonneux during a fierce German onslaught (1918).

TROOPER JAMES EDWARD MEYER

The story of Trooper James E. Meyer and his heroic adventures in the South African War is not yet finished because he decided to return and carry on the fight. For this purpose he enrolled with the Fifth Mounted Rifles Contingent that had been selecting men under the following conditions: — engagement to be for 12 months, or the duration of the war; preference to be given to those who had returned from active service in South Africa; single men to be preferred; pay to be at the rate of five shillings per day from date of joining the Camp of Instruction at Langwarrin; no man over 12 stone in weight was eligible; all must pass strict riding, shooting and medical tests; and Members of the Victorian Rangers were accorded preference over civilians.

James E. Meyer passed every one of these requirements and accordingly was accepted. The Fifth Contingent departed on 15th. February, 1901, with a total of 1,017 men and 1,099 horses, and three transports were necessary to convey this large Contingent to the war: — namely the ORIENT, the ARGUS and the CITY OF LINCOLN.

One must pause at this point to ask WHY did this brave survivor of the Elands River siege wish to return to the war, especially as he had almost died from enteric fever? And having displayed dashing feats of courage during his year with the Bushmen's Contingent, WHY did our hero feel he must return to the fray?

Like many other troopers with the Fifth Victorian Contingent, James took his own horse to South Africa where it probably soon died like thousands of other horses from shells, starvation, disease, or sheer exhaustion. The Fifth Victorian Contingent of Mounted Rifles arrived in South Africa, in 1901, at the start of the African winter which lasts from April to September. The biggest difference since James left there was the blockhouse line in process of construction.

Each blockhouse was a small, square building made of stone walls about nine feet high and with a corrugated roof. In the stone walls were loopholes for musketry fire, and each little blockhouse could hold from six to thirty guards. These stone forts were dotted along the railway lines, no more than 2,000 yards apart, and linked together with fences of barbed wire to limit the enemy's movements. Also, there were several armoured trains regularly patrolling the lines which helped to contain the enemy by stopping his nasty habits of blowing up the line and killing passengers. Within a few months Lord Kitchener's system of blockhouses produced the desired effect as many of the Boer guerrillas, finding it impossible to cross these lines of bullet-proof blockhouses, were being driven into a corner and captured by British mobile columns.

It was during this winter campaign that the 5th. Victorian Mounted Rifles were resting in camp on June 12th. at a place called Wilmansrust, situated between Middelburg and Bethel. After the long months of commando-chasing, the Victorians slept soundly, but in the darkness Viljoen's force descended heavily upon them. Surprised out of their sleep and unable to distinguish friend from foe, our men were shot down as they rushed to get their stacked rifles.

Within five minutes the Victorians had lost twenty killed and forty wounded and were so overwhelmed by superior numbers that they had no choice but to surrender and hand over all their supplies of food, clothing, ammunition and guns.

Trooper James Meyer was lucky to escape with his life; and it was to the Victorians' credit that they fought as well as they did under such impossible circumstances. After that encounter, Trooper James Meyer and a company of Mounted Rifles was operating in the Utrecht neighbourhood of the Transvaal when he had another narrow escape. He wrote about it to his widowed mother: -

De Jaeger's Drift — 24th. September, 1901. *"No doubt you have seen in the papers of the severe fight we had at Umvolosi River. I will try and give you a short description of our march and fight.*

Thirty men left our camp on the evening of the 25th. of July. Seven miles out we joined forty men from General Symon's Corps and arrived at Bamba Kop at 12-30 a.m. We picketted our horses in a large kraal and waited there till 2-30 p.m. before travelling another 10 miles to join our column under Major Jervis Edwards.

That night we camped at a signalling station. We were up at 4-30 and had saddled our horses and left camp within the hour. We travelled ten miles and camped in a large Boer farmhouse and garden, and during that night our outpost sentries captured two Boers, but five others got away. We moved on next day, leaving behind the waggons and two of the guns in charge of our captain and sixty men.

We took the third gun with us and travelled another 28 miles. We reached General Johann Grobler's house by 8 p.m. (Grobler is a friend of Smuts and one of the senior Boer leaders.) Unfortunately, Grobler must have been warned of our approach for he managed to leave the house, but our men fired three shots at him. He galloped away and I have since heard that he was wounded.

By the time we reached the top of a kopje our gunners were retiring across the river. The Hussars and ourselves now looked down on the valley through which the Boers were galloping. Then the Hussars retired, leaving us alone on the kopje with orders to "HOLD IT AT ALL COSTS!" Meanwhile the Boers had crossed the valley and were now climbing a high kopje that overlooked the valley through which our men had to pass. From that high ridge they could fire down upon the river.

After holding the kopje for an hour -and-a-half, we received orders to retire. As we were crossing the river we felt a hail of bullets; one of our horses was shot in the leg and one of our men was shot in the thigh. Shortly afterwards another man received an explosive bullet in his arm and another had his horse shot dead beneath him.

Our only gun had taken up a position to cover our retreat, but as we got near the gun the Boers attacked it in earnest. They got within 300 yards of it and their firing was so severe that the poor gunners had to retire and their officer received orders to destroy the gun and leave it. But then ten of our company (Mounted Rifles) were told to stand by the gun and not to leave it, even if every man of us was shot down. Nice orders to receive!

It was 2-30 p.m. when our sergeant-major and sergeant were shot down and killed. The firing was fearfully heavy. We started to eat our rather late breakfast when we heard two shots in the distance and five minutes later we saw some of our company riding over the veldt for their lives, closely followed by about 130 Boers. Our big gun fired two shots which checked the Boers for a while, but then they broke up into several smaller groups and spread out in all directions to try and surround us.

Myself and two comrades were told to watch the river about 2,000 yards away where the Boers were now crossing. Another man from our group joined us and we began firing in earnest. While we were blazing away at the Boers, our gun was able to retire to a position two miles behind us.

Suddenly the Boers began to pepper us from behind and under their deadly cross-fire we tried to retreat, but when we reached our horses they were mad with fear. I was the last to get mounted as the brute I was riding would not let me mount her, so my chum bravely rode back under heavy fire and held her head while I got on. Then we made a wild gallop towards our gun which was firing all the time to cover our retreat.

The Hussars and eight of our Mounted Rifles rode back two miles and took up another position, whilst my party was sent a mile further on to take command of a kopje near the river; but we could see the Boers were determined to cut us off and take our gun away.

Major Jervis Edwards and four of his men, with a sergeant of the Hussars, left us to take up a higher position on the kopje. When they did not return by 3-30 p.m. Major Henderson came to my group and said he was very anxious about Major Edwards and called for four volunteers to go with him to look for the commander.

Sergeant Murchie, myself and another young fellow named Lewis, stepped out. We advanced up the kopje, but the firing was the heaviest of the day so we took what cover we could behind anthills. When we had gone about 70 yards we noticed two Boers on top of the next ridge, so Sergeant Murchie gave us orders to drop.

I hid behind an anthill and Murchie hid behind another one about five yards away. Just as we dropped down there came two bullets and I thought they had passed between us, so I turned to Murchie and said, "That was close!" To my horror I saw he had been hit and had fallen on his back. I called his name, asking where he had been hit, but he never spoke again, or groaned, or moved.

Major Henderson could still see the two Boers above us, so he fired two shots at them and then crawled back, telling us to advance when the firing eased a little. But the firing just got worse.

Lewis and myself crawled up to Murchie and found the poor fellow was shot through the heart. I took his ring, watch and money; and then we retired by rolling over anthills and crawling on our bellies while the Boers fired down upon us all the time.

We got back to the rocks completely exhausted. it was 4-15 p.m. and we thought it was all up for us. We did not realise that Lieutenant Rundle with twenty men had beaten the Boers in a race for the kopje that commanded the only road we could escape through. Apparently it was a close shave, for they only beat the Boers by 100 yards and the race was so close that the Boers cut off their last man and took him prisoner.

So our men galloped away to the nek as fast as their horses could go and placed our gun there. It could now fire a few shots into the Boers from its high position.

I asked Lieutenant Montgomery if I could go out and fetch Murchie's body in. He said, "Yes, and I will go with you." Another man volunteered and we started off and soon reached the place where poor Murchie was lying. The lieutenant turned to me and said, "I would like to carry on and try to find Major Edwards. He may only be wounded."

We reached Murchie's body and then climbed up further and found Major Jervis Edwards and his sergeant lying 10 yards apart, both dead. We had no ambulance waggons with us so we had to leave the bodies lying there. We also found the bodies of the two Boers on top of the ridge, lying beside a rifle and two large pools of blood.

We returned and then received orders to mount. By now the Boers were retiring, but they fired a few shots at us over their shoulders as they left us. So ended our fight after nine hours. The horses had a very rough time of it, having their saddles on for 46 hours, and most of that time we were in the saddle. My poor horse was badly knocked up. Our casualties were 6 killed, 2 wounded and 4 prisoners.

I think the Boers must have lost very heavily. The

brutes came back that night and stripped our dead of their clothes, for when our waggon went out next day to collect the bodies, all were naked. I was one of the volunteers to dig their graves and I could not help thinking of the poem, "The Burial of Sir John Moore".

It was to capture our gun that the Boers fought so determinedly as the rebels are desperately short of guns and ammunition, not to mention boots and clothes.

We leave here shortly to go as scouts for General Hildyard's column. I am pleased to say the colonials are receiving great praise everywhere."

No doubt Trooper Meyer had many more similar experiences before the time came for his departure. He left South Africa with his regiment on 3rd. April, 1902, just two months before the end of the war, so he certainly witnessed the final throes of that tragic affair.

During the campaign between January and April of 1902, James observed the Boer resistance slowly ebbing to its close as many hundreds of rebels were captured by British mobile columns, a success due entirely to Lord Kitchener's system of blockhouses. This system had been improved while James was fighting there, from those early blockhouses built 2,000 yards apart to the more efficient models constructed only 600 yards apart with impenetrable strands of barbed wire connecting them. One soldier pointed out that if his hat blew over the wire he would have to walk around the last blockhouse in the line to retrieve it, a distance involving hundreds of miles. It is no wonder that by the time James left South Africa the Boers had been brought to their knees by British perseverance.

PROFILE - JAMES EDWARD MEYER

James was born in Donald on 1st. March, 1873, the third child of Johann August Meyer and Charlotte Bennetto. James' father was born in Heist-Holstein, Germany, and arrived in Donald during the year 1863. Johann August Meyer was a successful businessman who became Donald's first storekeeper, first publican, first butcher, first auctioneer and first livestock agent. J. A. Meyer was also a founder member of the Donald Shire Council and a Justice of the Peace for many years until his death on 26th. May, 1897. In fact, Johann August Meyer is nowadays regarded as the true founder of Donald. When James Edward Meyer left school he worked as an accountant in the local branch of the Commercial Bank and in his leisure hours he played football and cricket and was considered a good sportsman. He was also a member of the Donald Rangers, an organisation that existed in Donald during the latter part of the nineteenth century for the purpose of recruiting a body of men to be trained in the defence of their country. In October, 1896, James decided to leave Donald and seek his fortune in Western Australia because the Gold Rush had just begun in that "Land of Promise". James knew several Donald men who had gone there to find gold and possibly he was persuaded by the story of George Smith, of Watchem, who had only been in Coolgardie three days when he dug up a gold nugget worth 40 pounds. In December, 1896, James Edward Meyer took twenty picked horses by boat from Melbourne to Perth hoping to sell them at a profit, but unfortunately that did not happen. Actually, James was very unlucky because the horses sold for less money than he expected, barely enough to cover the cost of transporting them. To make matters worse, during the voyage someone stole his box containing all his good clothes and valuables, including a handsome gold watch from his father. Then although he searched and searched, he could not find a job in Perth that was suited to his qualifications. At the beginning of 1897 James made his way north to Coolgardie where he found a job in one of the mine offices. However, he did not stay long in this famous gold town because in June he received word that his father had died, which necessitated his immediate return to Donald. So then he lived with his widowed mother and three siblings and continued his father's work as a stock-agent. He rejoined the local branch of the Victorian Rangers and when the Great Boer War began on the 11th. October, 1899, James Edward Meyer enlisted in the Third Contingent known as the Bushmen's. Enlistment papers state that he was "born on 1st. March, 1873, single, a stock-dealer living in Donald, a Wesleyan, and 5 feet-9 inches tall with a chest measurement of 41 inches". No. 470, Private James E. Meyer, of the Third Bushmen's Contingent, sailed from Melbourne on the 10th. March, 1900, on the "EURYALUS". He survived many skirmishes with the enemy, but caught Enteric Fever after drinking polluted water during the Elands River Siege in August, 1900, and was invalided home, arriving in Australia during December, 1900. Almost immediately he volunteered for another tour of duty in South Africa, this time enlisting in January, 1901, with the 5th. Victorian Mounted Rifles Contingent. A diary and several letters to his widowed mother from the war zone portray a keen soldier whose devotion to duty is utterly praiseworthy and whose courage on the battlefield is of the highest order. On his return from the Great Boer War he married a nurse, Marjorie Fraser. They had two daughters who were both born in Donald: — Charlotte (1907) and Constance Marjorie (1909). For several years James Meyer's wife was in charge of a private Nursing Home in Blair Street, Donald, and worked hard to make it a success. At the Donald Show in 1906, James won three First Prizes for poultry, eggs and his Irish Terrier; and in December, 1907, he won a prize for "the best fishing catch". In 1910, James' wife cooked a traditional HAGGIS for the very first supper of Donald's local Caledonian Society. Then James and his family left Donald to settle in Melbourne because the 1916 Melbourne Business Directory has an entry: "MEYER, Jas. E. Commercial Agent for Lascelles". His daughter, Constance, was a nurse in the Fairfield Isolation Hospital and died from an infection contracted in the course of her duty there. When James Meyer's only sister, Lena Meyer, died in 1937, her obituary indicated that James was still living in Melbourne and the following entry was found in the 1938 Victorian Directory: — "MEYER, Jas. E. 21, Linden Avenue, Heidelberg West".

THE FIFTH VICTORIAN CONTINGENT

Private JOHN HAMILTON, of Donald, sailed for South Africa in a military contingent consisting of 1017 men, including 46 officers. They set sail from Melbourne on the 15th. of February, 1901, in three troopships (the ORIENT, the ARGUS and the CITY OF LINCOLN) and returned home on the 25th. of April, 1902, after completing their twelve months of service. The men of this 5th. Victorian Contingent were selected for their riding and shooting skills, a fact supported by comments in Private HAMILTON'S letter.

2nd. April, 1901. *"I am at Sunnyside Camp, in Pretoria, at the moment. We got to Cape Town safely after a smooth voyage, but it was not very pleasant as there were too many troops on board the ORIENT. The constant movement of men in khaki made me think of an enormous ant-heap whenever I looked down on them from the top deck.*

We stayed a week at Maitland Camp and were allowed to visit Cape Town which was only four miles from our camp; the train fare was one penny.

It struck us as very peculiar the way in which we were viewed by one and all, and what a warm reception the civilians gave us. You would scarcely believe the high opinion in South Africa regarding Colonial Troops; and the way we are sought after by the English Commanding Officers out here because they prefer us (which leads to no end of rows between Tommy Atkins and ourselves).

We sailed from Cape Town to Port Elizabeth and then entrained for Pretoria (in Northern Transvaal) which took us five days and five nights to reach.

My train was made the "advance guard" for all the other trains on the same route; and so we pushed two trucks of ballast in front of our engines to "Stand Sam" if there were any explosives planted on the line.

Then we had to lie down with our rifles in our hands (pointing through loopholes in the side of the truck) as we passed through the Orange River Colony, for there were small commandos (groups of Boers) raiding in different parts through which we had to pass; and so we were warned that at any moment we might have to make ourselves useful and protect the four supply trains in our convoy.

The first part of that overland journey was one continuation of winding and curving railway — lines, passing around kopjes and dongas, some of the kopjes being so steep-sided that a man would scarcely dare to climb them.

The countryside varied very little until we neared the border of Natal where the train crossed long, narrow strips of plain (veldt) and on some of the surrounding kopjes hereabouts we saw British troops trying to drive their tricky enemy back into open country.

Later on we passed through Arundel, Colesberg,

Rensburg and other places where some hot fighting had taken place earlier. We also saw the memorable Coleskop, near Pink Hill, where our gallant Major Eddy and other Victorians are now resting, undisturbed.

After this we got into more open country and the veldt here had an abundance of grass and the cattle and horses were in good condition; with the exception of those animals which had been recently commandeered by our troops. These animals were being guarded by the infantry soldiers who are stationed in forts about every mile or so along the railway line to stop the Boers blowing it up.

As we travelled through the night we missed a great many sights, though some of my mates on the train had been through the area before with previous contingents and had fought their way through these same places and so were able to personally relate details of the battles that had occurred here.

From the Natal border onwards we could see little wooden crosses that had been erected by soldiers to mark the resting places of their fallen comrades. In many places I counted up to thirty graves; and seldom did we travel more than a couple of miles without passing the cross of some brave fellow who died fighting a desperate enemy. After all, the Boers are only trying to save their cherished homeland; and now it is the desire of every Boer in the field (we are told) to die rather than surrender to the British.

The large bridges across the Orange River have all been blown up by the Boers, but are now being replaced with splendid structures erected by our English engineers.

I was disappointed in my first sight of the famous Bloemfontein, one of the poorest — looking towns I have ever seen, possessing just a few tin houses and a large enclosure surrounded by high fences of barbed wire — this is a jail for Boer prisoners and there were about 800 Boers in this prison compound when we passed by."

Private HAMILTON'S train journey was certainly fraught with danger for it was this long single line of railway from Cape Town to Pretoria which gave Christian De Wet (the most audacious fighter of all the Boer leaders) many marvellous opportunities for carrying on guerilla warfare. The British defended this line of railway strongly — every station, bridge and culvert was guarded — but still there remained many miles which could not be under constant surveillance. And nothing was easier than for De Wet, with some of his followers, to swoop down from their hiding-places among the hills, or from some friendly farmhouse, under cover of darkness and attack one unguarded spot after another.

This was the kind of fighting that went on for many months and De Wet had everything in his favour. Time after time he was chased here and there by our mounted troops, but just like a fox he would disappear into his innumerable hiding-places scattered across the countryside. He continued to attack villages, capture convoys, burn trains, seize valuable supplies, cut telegraph wires and lay dynamite to tear up the line. Yet no matter how hard the British tried, De Wet always managed to slip through their fingers.

Private Jack HAMILTON was at Sunnyside Camp, Pretoria, on **2nd. April, 1901**:

"I have seen the farm and fine homestead of the notorious Boer Commander, Christian de Wet. His house is surrounded by a plantation of different trees and nestles snugly at the foot of a large kopje on the top of which he had planted three large guns and had deep trenches dug. The British now possess this property, but De Wet himself is still at large at the head of a commando about 2,000 strong; and with his expert knowledge of the country he is fighting in he can easily evade his pursuers.

I have also seen the wonderful machinery in use at the gold mines of Elandsfontein. These Rand mines are believed to be very rich and judging by the expensive equipment put there by the Boers there must still be unlimited wealth in them.

I have seen the residence of ex-President Kruger which is a micely-designed building of brick with a pretty flower garden surrounding it and a porch covered with lichens and other creepers. It is situated in the centre of Pretoria; and nearby is a splendid-looking statue of 'Oom Paul' made of marble, but it stands there unfinished.

We are going out at any moment now to march on Pietersburg where there are 20,000 Boers concentrated, so you can be sure it will be a hot go and the generals think it could be our last big fight.

Anyhow, we are determined to uphold the good name of our predecessors and all hands are restless for a go. I wish you were here too, old boy; we could do with a few more mates from Donald. **Percy King** is with us and **Ronald Muir** is O.K. Aurevoir old chap, remember me to all the Donald boys. Your old pal, Jack Hamilton."

On the 8th. April, Colonel Plumer's force rode into Pietersburg to find that the Boer commando had already fled from the town.

FARRIER-SERGEANT PERCY KING

Before the war, Percy King was employed as a blacksmith at Mr. Falla's business in Donald. He enlisted in the Fifth (Military Rifles) Contingent and describes some of his experiences in the following letter to his Donald friend, Mr. J. Bolden.

> "**Bronkhorst Spruit. 25/7/1901** — Dear Joe, No doubt you read a little of our doings in the papers, but what amuses me about the Australian newspapers is that they make their readers believe the Boers are being driven all over the veldt and taken by surprise every hour of the day — but no such thing really happens.
>
> You have to be up mighty early to catch Johnny Boer asleep! He is for ever on the watch and travels always in small groups of any number from three to twelve guerrillas. But if there is a good action going on, then 30 or 40 of them will appear in less than half-an-hour.
>
> The Boers have ways of signalling to each other by lighting fires; and as they can be seen from a great distance riding against the skyline on top of the hills, this particular way of moving about seems to draw the attention of their mates.

There are any number of British columns moving about all over the Transvaal trying to catch the Boers. I suppose my column is the weakest with only about 1200 men, but there are some columns that have 6,000. We are escorting large convoys of supplies, but the Boers easily slip around these long convoys and vanish into the hills.

We have lost a good number of men since being out on escort duty, and we (Victorian Mounted Rifles) have to do the hard work. We are spread out over other columns, too. We have just returned from a 14 days' trek around Rhenoster Kop and as this is our third time over the same track we all know it very well and could not get lost even in the dark.

This is a lovely country and good weather for outdoor work, so far at any rate, as we have only had one day's rain since landing here. If I like the summer as well as I like the winter I will remain here. Of course, Johnny Boer may alter my plans, but so far he has treated me very well. I am keeping in the best of health and ride out with my Company every day; and I have shot away enough lead to make type to set the "Donald Times" newspaper.

Yet I do not know if I have even touched a Boer as there are hundreds of them. But you have enough heroes in Donald with Peter Falla and Arthur Hornsby. I was pleased to hear they are both back safely, although I heard Peter lost his arm. I have seen the spot where he was wounded.

My Company so far has only had one men wounded and one killed — **Captain Chomley** is a great officer and always gets us out of a tight corner.

Ronald Muir is in my Company and as happy as a

sand-boy. He is as fat as butter and I think he resembles Andrew Muir now more than ever. I saw Jimmy Muir from Laen. He is in the Scottish Horse and he is in good health. In fact, we all are. I weigh 14 stone, but no wonder as I have no work to do. I have three assistant farriers and they are good fellows and do all the work.

We have 186 horses to keep shod, but all the iron shoes arrive here already turned and clipped. I have a nice little portable forge with tools and there is plenty of coal out here although it seems full of sulphur.

Most of us like this life very much. My tent mate is a chap from Stuart Mill and he is Company Sergeant-Major William James Knights."

A few months after Percy wrote this letter, Captain Chomley was recommended for the Victoria Cross after rescuing Corporal Cummins by bringing him out of the action under very heavy fire. Both Percy King and Ronald Muir were in Captain Chomley's Company so did they witness his brave deed?

PROFILE – PERCY KING

King was at one time employed in Mr. Falla's business as a blacksmith. Percy played in the Donald Brass Band and was a very capable musician as well as the band's secretary. In February, 1900, Hyland's Circus visited Donald and advertised for musicians to play in its band. No doubt the job appealed to Percy because he loved playing music and he also liked the idea of travelling around

Australia, so he joined the circus band and left Donald. One year later Percy enlisted in the Fifth Mounted Rifles Contingent as a farrier. This Fifth Contingent departed from Melbourne on 15th. February, 1901, consisting of 46 officers, 971 other ranks and 1,099 horses. After a few months in South Africa, Percy was promoted to the rank of Farrier-Sergeant. In a letter to his former boss in Donald, Percy wrote that he was enjoying his experiences and thinking of staying in that country at the end of his twelve months. As there is no report in the local paper of Percy's home-coming one can presume that he actually did remain there. In fact, there were eighty-three men struck off the Fifth Contingent's roll in South Africa which indicates that PERCY KING was not the only Australian looking for "greener fields and pastures new".

Farrier-Sergeant PERCY KING'S letter to Joe Bolden was written in a British army camp at Bronkhorst Spruit (Transvaal) and it was here twelve months earlier on the 13th. June, 1900, that a small party of our Australian Mounted Infantry, a force less than a hundred in number, gained a kopje which overlooked the Boer army and fired every cartridge which they possessed into the enemy throng with such accuracy that they killed many men and horses.

This successful action by our Australians was highly-praised by British historian, Arthur Conan Doyle, who recalled an earlier event at the same place in 1881 when the Boers attacked and slaughtered a small British garrison: -

> "*Bronkhorst Spruit, at which the retreating Boers were so badly mishandled by a small band of Australians, was that very same place at which nineteen years previously*

a British regiment had been taken by surprise and shot down. Who ever could have guessed the nationality of the brave men who would avenge that terrible slaughter?"

"THE WORST THING THAT EVER HAPPENED TO AUSTRALIA!"

Private WILLIAM ALBERT KEMMIS, a resident of Cope Cope, was welcomed back from the Boer War at an official gathering on the 20th. of May, 1902. The local Banyena hall was packed and Private Kemmis sat on the stage with his brother, Corporal F. J. Kemmis, who had arrived back from the war some months earlier. Sergeant-Major Knights and Lance/Corporal McLennan also accompanied the guest of honour onto the stage. Councillor Reseigh, representing the Kara Kara Shire, presented Private Kemmis with a handsome gold watch and a medal which bore the inscription: -
PRESENTED TO W. A. KEMMIS BY HIS FRIENDS ON HIS RETURN FROM ACTIVE SERVICE IN SOUTH AFRICA — 1902.

Albert said he was never so happy in his life as when he landed back at Cope Cope Railway Station and was met by his old friends. Then he gave the audience a brief account of the WILMANSRUST night-attack on the 12th. of June, 1901, in which he, as one of the Fifth Victorian Contingent, was involved. He described it as the worst thing that ever happened to Australia.

"It seems so hard for me to describe the facts here. Our Victorians were completely capsized and suffered a perfect pandemonium in a night — attack made upon us by the Boers. It was the Muller and Viljoens Commando who surprised us out of our sleep on the 12th. June. The Boers got the best of us and gave us a regular bad time.

The scene after five minutes was something fearful to look upon and I hope that I will never have to look upon the same thing again. I can only refer to it as a second "Majuba" and am thankful to say that I came out of it without a scratch. I am fully convinced that I was never meant to be shot and I know Providence preserved me.

It was something terrible as everybody all around me was shot. Our camp numbered about 300 strong and the Boers were reckoned to be between 500 and 600 strong. Within five minutes we lost twenty killed and forty wounded. There were also 97 horses killed which without doubt saved many of our men who were able to shelter behind the dead horses.

We were taken completely by surprise. Most of our men were rolled up in their blankets, resting after a hard day's ride. The Boers crept up to within 30 yards of our lines and lay quiet for about 25 minutes listening to our fellows talking around the fires.

As fierce and wily as a pack of wolves, they silently crept up between our pickets (guards) and we did not know they were there until they opened at point-blank range upon our sleeping men. To make matters worse, our rifles were stacked in a neat pile (that is a strict military tradition) so most of our troopers were shot down as they rushed to get their weapons.

The Boers took all our stores and clothing, and then happily rode away with their spoils; while we, the survivors, were left alone with our dead and dying."

Everyone in the audience would have been moved to tears by Albert's story for it is easy to imagine tired soldiers sleeping in their camp while their cunning enemies creep closer and closer, forming a deadly cordon around them; and no doubt his family and friends thought Albert was very lucky to survive that terrible slaughter.

Any blame for the tragedy that befell this small unit of Mounted Rifles must be aimed at their commander. He was negligent in failing to take special precautions to protect his men as he should have anticipated that their temporary camp would be exposed to danger from wandering Boer guerrillas.

Still, it is comforting to know that military records state our brave Australians did as well as any troops could have done who were placed in such an impossible situation. Surprised out of their sleep, without their rifles, and unable to distinguish friend from foe in the darkness, it was a miracle that anyone survived, but Albert Kemmis did and returned home to tell us all about it.

PROFILE – WILLIAM ALBERT KEMMIS

Kemmis was born in 1865, the son of Avon Plains settlers, Henry Kemmis and Ann Smith. He grew up on his parents' selection and worked hard. He volunteered to fight in the South African War and consequently enrolled with the Fifth (Mounted Rifles) Contingent which left

Australian shores on 15th. February, 1901. On arrival at Pretoria, Albert Kemmis soon found himself chasing Boer commandos who were usually invisible and therefore impossible to catch. Somehow Private Albert Kemmis survived many encounters with these Boer guerrillas and safely returned to his parents' farm where he spent the rest of his life as a successful grazier. He never married, but was a great community worker: — a member of the Avon Plains Rifle Club; a member of the Barnaby Lodge of the Masonic Fraternity; a member of the Avon Plains School Committee; and a Board Member of the St. Arnaud Hospital. Mr. H. W. Kemmis paid the following tribute to his brother:

"*No straighter man than my brother, Mr. Albert Kemmis, has ever lived. His word was indeed his bond, and he was ever ready to do a good turn. We lived and worked together at Avon Plains for 76 years, and the only time we were parted was when Albert went to the South African War with the 5th. Victorian Rifles. Had he still been alive here now, the farm would not have been sold. He always tried to help the district in any way he could. For example, once when the local school was to be closed because the parents could not afford to pay a teacher, Albert, though having no family himself, deemed it a privilege to contribute. His name has always been connected with the local hospital and he was one of their best practical supporters.*"

The announcement of his death in the Donald Times simply said: "William Albert Kemmis, an old and prominent resident of Avon Plains, died after a long illness on 24th. February, 1940, aged 74 years."

PROFILE – PRIVATE JOHN HAMILTON

Hamilton left Donald with his friends, Percy King and Ronald Thomas Muir, but there is no record of his return. In the Fifth Contingent's Nominal Roll, Private John Melrose HAMILTON, No. 955, is reported to be dangerously ill at Kozi River on the 28th. September, 1901.

Two other Donald district soldiers who enlisted with the Fifth (Mounted Rifles) Contingent were REGINALD PHILLIPS, who went as a Shoeing-Smith, and FRANCIS JOSEPH CHAMBERS. Both men were sons of Avon Plains settlers.

PROFILE – JAMES DUNCAN

Duncan left Melbourne with the Fourth (Mounted Rifles) Victorian Contingent in which there were several other Donald men. On May 1st. 1900, he sailed for South Africa on the "VICTORIAN" and without doubt he saw plenty of action as a Mounted Rifleman. His gallantry at Wolmarananstad on 6th. March, 1901, earned him the M.I.D. and later he was promoted to Lance-Corporal. He also served his country in the First World War with the same courage and unselfishness and died in an English hospital during May, 1917, after being severely wounded in action in France.

PROFILE – DUNCAN JAMES McLENNAN

Duncan was a bushman from Banyena when he enlisted with the Fourth Contingent. His father was John McLennan, a farmer of Banyena. Duncan was nearly 5ft.-10ins. tall with a 41 ins. chest, a fine specimen of a country lad who could ride any horse and shoot straight. He enrolled as a Bugler and later was promoted to Lance-Corporal. He was invalided back to Australia, arriving home in Donald during August, 1901.

PROFILE – JAMES EDWARD HOLLAND

Another Donald young man who went to South Africa was JAMES EDWARD HOLLAND, a grandson of the well-known Avon Plains pioneer, Mr. John Holland. James was educated at the Avon Plains School and then assisted his father on the farm. He volunteered for active service in 1900 and no doubt risked his life many times while fighting with the Fourth Victorian Contingent. He was promoted to Corporal during his period of service and it is possible that he stayed in South Africa after the war.

Trooper Richard Merrett

WOUNDED AT ROOIWAL

As soon as he was eighteen years of age RICHARD MERRETT, of Massey, tried to join the last Victorian Contingent going to South Africa. He was rejected because at that time preference was being given to returned soldiers who wanted a second tour of duty. However, our would-be soldier was not to be discouraged so paid his own passage to South Africa. His boat arrived in Durban at the beginning of March, 1901, and he wrote home: -

> "I am off to the front tomorrow, 9th. March. I joined the Scottish Horse for twelve months at 35 shillings a week, all found. I leave at 5 a.m. tomorrow to travel through Natal (via Ladysmith) and into the Transvaal where we commence scouting duties right away.
>
> "I'm in Tullibardine's Cavalry which is mostly colonials. Durban is a nice place, situated amongst the hills, with a pretty bay. They have got electric light and sewerage here and the blacks do all the labour. It is amusing to see a native pulling his rickshaw containing two or three heavy passengers."

Trooper RICHARD MERRETT's job was patrolling the

railway line at night to stop any Boers from blowing it up. This part of the Transvaal was the hiding-place of Delarey, a formidable Boer general who had managed many times to outwit the British columns forever chasing him.

30/3/1901 *"I have seen the Boers when I have been on patrol duty, but they always keep a thousand yards off. One of our patrols got cut up the other day and one trooper was wounded and three horses killed. The Boers fired at the wounded man on the ground, then they came up to him, took his rifle and the saddles from three dead horses and all the ammunition, and then told him he could go. He got up and walked the seven miles back into camp.*

There are plenty of Boers around Zanspruit and it is not safe on patrol as you don't know when the bullets will come at you. We have horse races after pay-day and they are very exciting, with our best riders competing.

I live in an iron house, not a tent, but there are plenty of rats about here. The other day one of our cooks got drunk and shot a Red Cross man and the poor chap died 20 hours later. The cook is going to be tried by court-martial for manslaughter and if guilty he will be shot."

When Trooper MERRETT wrote to his father in March, 1902, he told the amazing story of the capture of their great general, Lord Methuen: -

"At the 'BATTLE OF TWEEBOSCH' Lord Methuen was wounded in the groin and they tell us there is a great danger of him losing his leg. It happened like

this: — early on the morning of March 7th. a thousand riflemen belonging to Delarey charged straight at Methuen's convoy, firing from their saddles as they attacked.

There were many British infantrymen wounded and killed and over 200 prisoners taken, including our grand old General. Convoy, ammunition, guns, horses — the Boers took everything and it was a great victory for them.

However, the Boers were considerate to all their prisoners, and Delarey did not actually keep Lord Methuen very long but sent him back to his own lines with a Boer doctor to assist him. They say the British now feel greatly indebted to Delarey for his chivalrous behaviour at returning their important general to them."

Trooper RICHARD MERRETT was badly wounded in one of the very last skirmishes of the war on the 11th. of April, 1902, at a place called ROOIWAL. He wrote to his sister, Alice, from the hospital in Elandsfontein: -

"I suppose you heard I got a bullet in my top piece? I am getting along alright now but I felt a bit sick for a few days and am a bit deaf in the right ear.

The side of my face is swollen, but I have no pain in it now, only when the doctor is pulling me about. It was a good job for me it was a clean bullet. When I got hit it stunned me for about a day.

I got your letter on the 14th. of this month and you wanted to know if I had any been in any more fights lately. Well, I have had one battle too many and I don't want any more like the last!

> It occurred on the morning of April 11th. at a place called ROOIWAL when we were with Colonel Kekewich's column on our way back towards Klerksdorp. Suddenly, fifteen hundred of the enemy fell upon our column and we were taken completely by surprise. It was a terrible shock as I could have sworn there was not a single Boer within twenty miles of us.
>
> I never saw a more splendid attack. The Boers charged at us in one distinct line, four deep, knee to knee. We stood our ground and poured a steady fire into their advancing wave of horsemen, but they came on regardless.
>
> Unfortunately for me, their cavalry charge was well-covered by the fire of other Boers who had taken up positions on a nearby hill — and that is how I got a bullet in my head. I don't remember anything after that.
>
> My good mate, Holt, does not like South Africa and Harry Blair wishes he had never seen the place. You need not write after you get this as my time will be up on the 6th. July and I will not do any more time in the Scottish Horse."

Trooper RICHARD MERRETT was wounded in one of the last battles of the war, proving himself to be a soldier as good as any, prepared to fight to the death. On May 31st. Boer leaders signed a Peace Treaty at Vereeniging agreeing to Great Britain's generous terms of surrender; and it was our bushmen from the back blocks of Australia who played a major part in this great victory.

PROFILE – TROOPER RICHARD ACKLAND MERRETT

Merrett was born in 1882, the eldest son of Richard Henry Merrett and Elizabeth Ackland, of Massey. He worked on his father's farm until the outbreak of war between Great Britain and the Dutch Boers of the Transvaal and the Orange Free State. Unfortunately he was only seventeen and much too young to be selected into the Victorian Contingents, but nothing daunted he paid his own passage to South Africa where he immediately joined the Marquis of Tullibardine's 2nd. Scottish Light Horse — one requirement for this corps was that the men had to be Scottish or of Scottish descent. Consequently, Trooper R. A. Merrett served from 6/3/1901 to 24/6/1902 and when enlisting he gave his age as 19, his profession as a farmer, and his religion as Presbyterian. Trooper Merrett served in the final stages of the Boer War which in some ways was the most dangerous and frustrating part because the British were fighting an "invisible enemy". Trooper Merrett was wounded at Rooiwal, and soon afterwards Richard's father received a kind letter from **Lady Katherine Tullibardine** telling him about her visit to his wounded son and assuring Mr. Merrett, in the kindest manner, that there was nothing serious to worry about. After the war Richard went to New Zealand to work for about two years, and then returned home to his family's farm at Massey. In 1906 he married Miss Catherine McKinnon at Carron and continued to live at Massey where they shared the hardships of pioneer farming and brought up a family of six. Mr. Richard Merrett was a prominent, worthy citizen of that rural community and laboured hard on his farm at wheat-growing and sheep-raising. He was on the Massey School Committee, helped to run the local post office, a member of the Church of England Vestry, and also a member of the Masonic Lodge, the Donald Rifle Club and various other bodies. He died on the 10th. November, 1956, at the age of 74 and was buried at Watchem Cemetery.

N.B. The reader will perhaps have noticed that Trooper Richard Merrett's last letter was addressed to his sister in which he tells her not to write again because he would be seeing her soon. That reunion never took place because Alice Merrett died, only 21 years old [on the 23rd. of June, 1902 at Massey just prior to her brother's home-coming. Was this tragedy the reason for Richard's immediate departure to New Zealand where he worked for a couple of years?

A GENTLEMAN OF THE ROAD

Lord Methuen was the only British General captured during the Great Boer War and this contemporary cartoon cleverly illustrates that unfortunate event.

The capture of Britain's important general, Lord Methuen, was described in Trooper Richard Merrett's letter. At Tweebosch, on 7th. March, 1902, a large British column was suddenly surrounded by Delarey and his burghers. At the time, Methuen's horsemen were inexperienced soldiers so the sight of a large body of Boers charging straight at them, five lines deep and firing from their saddles, was so terrifying that they turned their horses and fled, leaving the brave gunners and Northumberland Fusiliers to hold the Boers off from their prey. It was a hopeless situation, but Delarey and his men showed kindness and courtesy towards the 121 wounded and 205 prisoners — **and even returned Lord Methuen**.

Private Henry Blair

Henry Blair and Charles Routley

THE SECOND BATTALION AUSTRALIAN COMMONWEALTH HORSE

CHARLES THOMAS ROUTLEY was born in 1880 at Donald East, the son of early settlers. From an early age Charles had to help his widowed mother with all the farming activities and therefore could ride and handle horses, working with them in all situations on the land. He enlisted in January, 1901, and fought in South Africa with the Fifth Contingent. He returned safely in January, 1902, and then enlisted again and was accepted because of his previous experience. This time he served with the Second Australian Commonwealth Horse.

HARRY BLAIR was born in Donald in 1878, the sixth son of Robert and Sarah Ann Blair who had twelve children. His parents were early settlers at Vigical, near Watchem, so Harry had plenty of experience working with horses on the land, as well as using horses to dig dams and grade the roads. Harry travelled to Tasmania, Western Australia and America before volunteering for active service in South Africa.

These two chums, CHARLES THOMAS ROUTLEY and HENRY BLAIR, joined the Second Battalion Australian Commonwealth Horse. This regiment gave preference to men with previous war experience, although candidates were also accepted who were acquainted with country life and with the management of horses — but they had to be good shots. Married men were not selected, except for N.C.O.'s of exceptional merit. Soldiers were supplied with rifles and bayonets and bandoleer belts, also a horse and saddlery.

Privates like Routley and Blair were paid 5 shillings a day; Corporals, Lance/Corporals and Buglers were paid 6 shillings; while Lieutenants were paid 15 shillings. The 2nd. Australian Commonwealth Horse left Melbourne on the 12th. February, 1902, on the transport "ST. ANDREW", with 21 officers, 351 other ranks and 401 horses. They arrived in Durban on the 10th. March.

From Durban they entrained for Newcastle where they were inspected by Lord Kitchener; and on the 23rd. April they moved out under General Ian Hamilton to begin a long trek which involved a series of "drives". Their aim was to push roaming commandos against the blockhouse lines, thus placing the enemy like meat in a sandwich. However, occasionally the cunning Boers would slip through British lines by wearing stolen khaki uniforms, an old and favourite trick.

On the 11th. May, Privates Routley and Blair took part in a big haul of prisoners when Commandant Bazendekout and his commando surrendered under the white flag. The British haul was 191 prisoners and over 200 horses. Then before arriving back at Vryburg their contingent took 52 more prisoners, including Jan Delarey, brother of the famous Boer general. The "drives" continued until eventually the Boer rebels were so harried and weakened that they were ready to seek peace.

Privates ROUTLEY and BLAIR returned to Klerksdorp on the 21st. May and remained there during the Peace Conference. Some time after the war ended on the 31st. May, their battalion was marched to Elandsfontein where they handed in their horses, saddlery and rifles, and later left there and entrained for the port of Durban. They boarded the "NORFOLK" on the 8th. July and arrived safely back in Melbourne on the 2nd. of August when they were paid off and discharged that same day.

So ended the war of CHARLES ROUTLEY and HENRY BLAIR. Their few months in South Africa had been a great experience, but very hard work. They had spent long, exhausting hours in the saddle, searching remote areas of the Transvaal, and "mopping up" small groups of Boers. It was very dangerous work because at any time a sniper might put a Mauser bullet in one's back. But worst of all, on their long "drives" across that dry and dusty land they had very little to eat and even less to drink.

In the same contingent as CHARLES ROUTLEY and HARRY BLAIR was Lieutenant JAMES POWELL who had previously fought in South Africa with the 4th. Victorian Contingent. With him was his younger brother, Private WILLIAM POWELL, going to war for the first time. These extracts from William's diary give readers some idea of what it was like to be there: -

> **APRIL: 6th**. *Broke camp at 12-30 p.m. and marched to Ingogo River. No blankets, so had to sleep in our overcoats. We are close to Majuba Hill where the Boers wiped out four hundred British Infantrymen in 1880.* **7th**. *Had swim in Ingogo River and spent the afternoon*

fishing and washing. **8th.** *Broke camp at 6-30 a.m. We passed right by Majuba Hill and saw graves of General Colley and all his company so I plucked a flower from his grave.* **9th.** *Arrived Elandsfontein at 5-15 and had a swim. Left for Klerksdorp at 7 a.m. Boer Generals, Botha, Steyn, Delarey and De Wet are reported to be here in conference with Lord Kitchener.* **10th.** *Captain Martin on Kitchener's staff inspected our camp. I was on stable-picket.* **11th.** *Went riding into Klerksdorp with brother Jim and had a look through the cemetery at all the soldiers' graves. Also we saw a Boer concentration camp.* **12th.** *Heard today that two trains with New Zealanders on board collided and many killed and injured in the disaster. It poured all night and we were soaked.* **13th.** *Marched about five miles from camp to funeral of dead New Zealanders. Over 2,000 present.* **20th.** *On Mounted Police duty in Klerksdorp as this is our Transvaal base for operations against Delarey.* **22nd.** *Lord Kitchener inspected our camp. I spent all day chipping lice off Lieut. Germaine's horses, then I went on stable-picket.* **23rd.** *We begin a long trek today under General Ian Hamilton, so broke camp this morning at 8 a.m. with a thousand men in our column.* **24th.** *Broke camp at 4 a.m. and passed through Haartebeestefontein town at 9 a.m. We picked oranges galore, then camped at Plaate Flaat.* **25th.** *Busy digging trenches. Captured two Boer spies, then spent all day destroying mealies which are the maize crops planted by the Boer farmers. It is the main source of food for the Boer commandos.* **26th.** *Broke camp and moved out at 7 a.m. Burning the dry mealie crops again, also burnt Boer waggons. Saw any amount of deer on the march.* **27th.**

Destroying mealie crops again and dug up dead Boers to re-bury them properly. Plenty of partridge and quail here. We commandeered 25 goats and 1 sheep from a Boer farm. **28th**. Broke camp and then marched about two miles. Washing and swimming all day. **29th**. Blew up a Boer house where commandos could hide. Burnt a coffin. **30th**. We had a camp fire and concert in the evening. I had my saddle stolen. **MAY: 1st**. Broke camp at 8-45 a.m. Lieutenant Cahill opened fire on a group of Western Australians by mistake as he thought they were Boers. **2nd**. Lieutenant Germaine and his group captured a Boer, after a 3-mile gallop. Burning mealies again today. **3rd**. Destroying mealies. **4th**. Arrived back at Kameelfontein. Convoy went to fetch rations. Boer prisoner sent on to Klerksdorp. **5th**. Out with General De Lisle on patrol and gleaned a few prisoners in Delarey's country again. Had a hard days's galloping. **6th**. Went up to the concentration camp and had a talk with the Boer women. **7th**. Had two falls from my horse today, but we start a big "Drive" against the Boers. **8th**. Broke camp at 7-30 a.m. and camped that afternoon in a pleasant valley. We captured four Boers. I was on outpost duty all night. **9th**. Broke camp at 7-30 a.m. Commandeered cattle and sheep. We camped near the railway and telegraph lines and saw eight Boers in the distance. **10th**. Broke camp at 7-30 a.m. Captured 2 waggons, 20 donkeys and several head of cattle and sheep. Eighteen Boers reported just ahead. Camped on border of Transvaal and Bechuanaland. Privates Bradford and Shaw on observation post were surrounded by Boers. Bradford was captured, but Shaw escaped with two bullets in his horse's breast. We expected an attack at any moment. Then a party of about 250 Boers tried to break through our blockhouse lines, but they failed. **11th**. This was a big day for us. During the "drive" we took 251 prisoners, 144 rifles, 300 horses, 6,000 rounds of ammunition, 450 cattle, 8 waggons, 700 sheep, 40 donkeys and 21 mules to send back to our Brigade. In the afternoon Boers came into our camp waving the white flag of truce; then 197 of them came to surrender and shortly afterwards 51 more arrived to surrender. **12th**. Broke camp. Colonel De Lisle left for England. We shifted about 7 miles and joined Colonel Williams' column. Brother Jim went into hospital. I got Botha's bandoleer. **14th**. Scarcity of water. Searching for water all day. **16th**. Found a half-caste, shot twice through the chest and neck by the Boers as they show no mercy to any native who is found fighting for the British. **17th**. Still chasing Mr. Delarey. **18th**. Today we camped at 2-30 p.m. on the battlefield at Rooiwal where 1,900 Boers attacked General Kekewich's column last week (11th. April). We heard that they suddenly appeared and charged a British column whose men were taken by surprise but bravely stood their ground as the enemy galloped towards them in one long line, firing from the hip. Our men coolly returned a steady fire upon the advancing wave of enemy horsemen until they turned and fled. The British lost 50, but the Boers' loss was much greater. I saw the fierce Potgieter's grave who used to beat his men with a rhinoceros-hide whip (sjambok) to make them fight. The toes and fingers of dead Boers were showing above the ground where sixty Boers had been buried rather hurriedly, so we had to dig proper trenches and bury them. **20th**. Broke camp at 7-15 a.m.

and arrived near Harts River where the Canadians were cut up badly nearly three weeks ago. The 2nd. Canadian Mounted Infantry were in Colonel Cookson's column and while chasing an army of Boers they found themselves cut off and overwhelmed by superior numbers. Eighteen out of twenty-eight Canadians were hit as they fought bravely beneath a perfect hail of bullets. One of the bravest deeds of this war. **June. 1st**. There was great rejoicing when General Ian Hamilton read out that Peace had just been proclaimed. Major Brown received the D.S.O. medal and then we had a football match between New South Wales and Victoria and we won. **2nd**. Each man was given 1/6d. worth of wood, free in honour of Peace. **3rd**. Troops are being urged to fight in the Boxer rebellion in China. **4th**. Warned at 2-30 p.m. that **I had been chosen to represent Australia at the Thanksgiving Service in Pretoria**, so had to leave camp at 5-30 p.m. for Klerksdorp. **5th**. Left Klerksdorp in a coal-truck for Pretoria.

In August, 1902, the soldiers of the Second Australian Commonwealth Horse returned home having upheld the fine reputation of those who had gone before them and well-deserving of this praise from Arthur Conan Doyle: -

"The newly — arrived Australian Commonwealth Regiments gave a brilliant start to the military history of their united country by the energy of their marching and the thoroughness of their entrenching."

PRIVATE HENRY BLAIR returned from South Africa in

August, 1902, and worked on his parents' farm at Watchem. When World War 1 began he unselfishly answered the 'call of his nation', although he had to convince the Army that he was fit enough at forty. While serving overseas he married a Cornish girl, Ivie Williams, and brought her back to Watchem where they farmed and raised six children. Harry died in 1936 and present at his funeral were Watchem's two surviving South African War veterans, Richard Merrett and William Walder.

PRIVATE WILLIAM POWELL resumed farming activities at Sheep Hills and married Catherine Bourke, of Carron, and they had four children. William took a great interest in community affairs. He became a hard-working Shire Councillor and dedicated himself to the service of the community with the same self-sacrifice that was apparent in his career as a soldier of the British Empire. He died in 1954.

LIEUTENANT JAMES POWELL returned to Sheep Hills in August, 1902, and later married. He served his country and empire again in the First World War and on his return from that terrible struggle he lived in Melbourne until his death in 1955. His stomach troubles in later life were attributed to drinking contaminated water during the South African campaign. He said that men were desperate and had no choice but to drink from dirty pools; so in order to find cleaner water they would push a hollow grass-stem below the water's filthy surface and suck up the water from underneath. No wonder more soldiers died from typhoid fever than from bullets. Lack of water was their greatest problem when trekking across the African veldt.

WILLIAM
POWELL
of SHEEP HILLS

WILLIAM
THOMAS BANDY
of CORACK EAST

FOURTH BATTALION COMMONWEALTH HORSE

WILLIAM THOMAS BANDY was born in 1882, the eldest child of Thomas and Mary Bandy who owned "Pine Forest" farm at Corack East. From an early age William worked on the farm and became skilled in the care of horses. He was a keen member of the Corack Rifle Club and was an excellent rifleman. When war began in 1899 William was too young, but he volunteered for active service in February, 1902. He left Donald to enlist in Melbourne, hoping to enrol with the Second Battalion Australian Commonwealth Horse; but there were too many returned soldiers wishing to re-enlist and these veterans had preference over all other volunteers. William passed the necessary examinations with flying colours and even returned to Donald to bid farewell to family and friends, but the battalion had filled its quota so William had to wait. Then Private William Bandy (No. 2268) enrolled with the Fourth Battalion Australian Commonwealth Horse.

On the 26th. March, 1902, William Bandy left Melbourne on the transport "TEMPLEMORE" which picked up troops in Adelaide and Freemantle, then arrived in Durban on the 22nd. April. The troops disembarked the following morning and camped in the showgrounds. Then on the 24th. April they entrained for Natal and after a two days' journey arrived in Newcastle and made camp at the base of Kitchener's Kop on the veldt, about three miles from town. Here they were inspected by General Burn-Murdoch, although by this time PEACE talks were in progress. However, on the 10th. of May they were ordered to Wool's Bridge, Buffalo River, just to make a show of strength at the bridge where the Boer delegates would be crossing on their way to the conference in Vereeniging.

On the 13th. May the soldiers were marched to Volksrust, there to entrain for Elandsfontein where the battalion remained until the 24th. June, busily engaged in re-fitting, sending sick horses to sick horse farms, and looking after newly-arrived remounts. No doubt Private William Bandy enjoyed this work with sick horses and was able to use his knowledge and experience to care for them. On the 28th. of May, General Oliphant inspected and complimented the men on their work with the horses. PEACE was proclaimed on the 1st. June and there was rejoicing in camp with a special announcement that every N.C.O. and Private was to receive 1/6d. worth of goods FREE from the Field Force canteens — a concession to troops that cost Britain £20,000.

William Bandy's family and friends happily welcomed him home with a concert in his honour at the Corack Mechanics' Institute on the 19th. August, 1902.

THE HERO'S RETURN

Below is a typical Victorian drawing-room picture of the Boer War period which tells a soldier's story far more effectively than words could. Appearing in homes throughout the nation, such pictures caused a genuine uprush of public concern for the sick and wounded soldiers arriving back from the war. As well as pictures like "The Hero's Return", for the first time in history **photographs** were taken that exposed the real facts of fighting with all its bloody butchery and brutality.

THE TREATY OF PEACE — 31ST. MAY, 1902.

After two years and seven months of hostilities, a conference opened at Vereeniging and 150 Boer leaders met together to discuss terms of peace with Britain. The guerrilla leaders were invited — Botha, De Wet, Delarey, Lucas Meyer, Beyers, Kemp and Muller — who all put their names to the treaty. At last there would be peace in the whole of South Africa, from Cape Town to the Zambesi River. However, a big mistake (No. 8) was that Britain entrusted the future welfare of South Africa's native population into the hands of the Boers. The Vereeniging treaty should have stated that "natives must be given the franchise".

TERMS OF THE TREATY: — 1. That the burghers lay down their arms and acknowledge themselves subjects of Edward VII. 2. That all Boer prisoners willing to take the oath of allegiance will be returned to their homeland. 3. That their liberty and property be inviolate. 4. That an amnesty be granted. 5. That the Dutch language be allowed in schools and law-courts. 6. That rifles be allowed if registered. 7. That self-government be granted as soon as possible. 8. That no franchise be granted for natives until after self-government. 9. That no special land-tax be levied. 10. That the people be helped to re-occupy the farms. 11. That three million pounds be given to help the farmers. 12. That the rebels be disfranchised and their leaders tried, but no death penalty.

DID YOU KNOW THAT....?

Colonel Baden-Powell who commanded the British garrison during the siege of Mafeking may have been the first man in history to think of creating a mine-field as a trap to destroy the enemy. He buried lyddite in a field outside the town and then his men withdraw. When the enemy followed, the mines were exploded and 1500 of the enemy were killed, or wounded.

Mr. Cecil Rhodes, the richest man in South Africa, left money for educational purposes. In his will he left six million pounds, out of which two million pounds were to pay for scholarships to Oxford University. His will stated that sixty colonial scholarships were to be filled annually by male students from various places belonging to the British Empire; including Rhodesia, Natal, Victoria, New South Wales, Queensland, South Australia, Western Australia, Tasmania, New Zealand, Ontario, Quebec, Newfoundland, the Bermudas and Jamaica.

The following dates of inventions are connected with Queen Victoria's reign: — on the day of Queen Victoria's birth — 24th. May, 1819 — the first steamboat crossed the Atlantic Ocean: in 1825 when she was six years old, the first railway train in the world started to carry passengers: in

1837 when she was eighteen years old, the Morse system of telegraphy was first used: and in 1875 when she was fifty-six years old, the first telephone was used.

During the war a particularly nasty act of treachery was committed at Warmead, a railway station north of Pretoria. Two British officers of the intelligence department named STEERE and KERR were talking to some Boer leaders who had said they were desirous of surrendering when other Boers who were hiding nearby fired at the officers and shot them dead.

The Modder River, whose contaminated water was eagerly drunk by thirsty troops, was nicknamed "Chateau Modder". It was certainly a full-bodied wine holding as it did many corpses, both Boer and British. Its corpse-laden waters contained cholera and dysentery germs, providing a lethal cocktail to those who drank it and killing more British soldiers than the Boers did.

The Great Boer War began at 5 p.m. on the 11th. October, 1899, and ended with the Treaty of Vereeniging on the 31st. May, 1902. It was Australia's first national involvement and about 17,000 Australians signed up to go from here to fight in South Africa. Six of them became the first Australians to win the Victoria Cross. Appropriately, all six Victoria Crosses were awarded for rescuing wounded soldiers under fire.

When Lord Roberts and his army halted at Bloemfontein for six weeks from March 13th. to May 1st. (1900) the greatest misfortune of the war occurred, for there was a great outbreak of enteric fever among his troops. One hospital alone was packed with 1700 soldiers suffering from enteric fever; and other hospitals were just as crowded. As many as fifty men died from it in one day, and more than 1,000 new graves in the cemetery appeared during that time. Bad water cost the British more deaths than the bullets of the enemy.

It was in this war that we saw for the first time the use of refugee/concentration camps, trench warfare, observation balloons, armoured trains, guerrilla warfare, blockhouse lines, landmines, and a scorched-earth policy. All of these would be used again in the following wars throughout the 20th. Century.

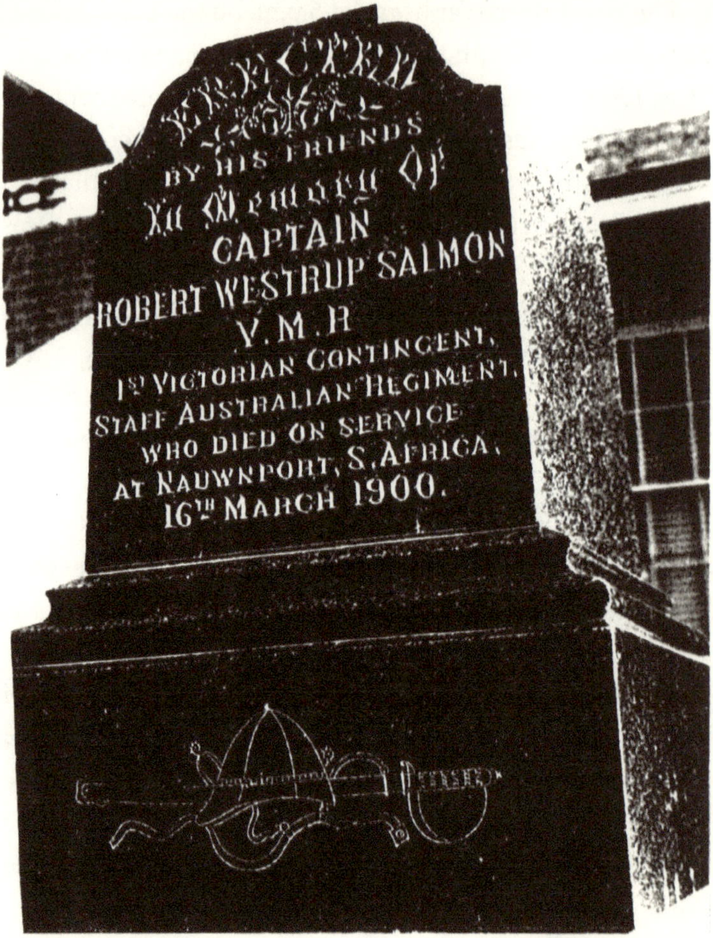

> **By his friends**
> **In Memory of**
> **CAPTAIN**
> **ROBERT WESTRUP SALMON**
> **V.M.R**
> **1st VICTORIAN CONTINGENT,**
> **STAFF AUSTRALIAN REGIMENT,**
> **WHO DIED ON SERVICE**
> **AT NAUWPORT, S. AFRICA,**
> **16th MARCH 1900.**

Captain Salmon died of enteric fever.
Trooper Arthur Horns by was his servant/batman
Memorial situated outside Talbot Post office
Photo: D. Breed.

Midshipman Huddart.
the first Australian killed in the Boer War

SIR REDVERS BULLER was the most under-estimated British general of this Great Boer War and deserves better recognition for his efforts. He came from Devon, a place famous for its naval heroes, such as Sir Francis Drake, but less noted for its military men. Buller (1839-1908) joined the army at 18 and served in China and Canada before the Zulu War broke out in 1879. In this war the small British forces were often desperately outnumbered by the fierce Zulu tribes and there were many acts of British heroism in tight situations. Buller was awarded the VICTORIA CROSS for valour when he led his 1,800 men from Inhlobane whilst hotly pursued by 20,000 Zulu warriors.

Some folks might argue the Great Boer War was really started by a humble English carpenter named GEORGE WALKER, for it was in 1886 whilst digging the foundations to build a small cottage that Walker's shovel uncovered a clear gold streak. His discovery immediately brought geologists to the area and they soon declared that the whole Witwatersrand ridge (in the Boer Republic) could be one massive field of gold; and thus began a Gold Rush which brought thousands of money-seeking foreigners (Uitlanders) into the Transvaal

When thousands of our Australian Bushmen landed at the Port of Biera, instead of going north to the front they were kept waiting for weeks in that unhealthy district where fever and dysentery undermined the constitutions of hundreds of them. An English journalist wrote: *"To say they were extremely annoyed would be describing their feelings too mildly. They were very savage, they forgot themselves slightly and swore with force and originality. They cursed Biera, they cursed their fate, they cursed various governments, but mostly they cursed their own government, for they were very political people, these Australians."*

During the siege of Mafeking, Colonel BADEN-POWELL insisted on keeping up his men's morale by organising frequent dances, concerts, sports, football and cricket matches. Eventually, one of his officers declared he would rather face the Boers' bullets than have to sit and listen to another of B.P.'s comic recitations.

This year, 1999, Australia spent $1,500 refurbishing the grave of HARRY 'BREAKER' MORANT, our famous poet-soldier who was executed for war crimes during the Boer War. Apparently Morant's grave in Pretoria had been vandalised and was showing signs of age; so now his grave (which also contains fellow soldier Peter Handcock) has been spruced up with a new concrete slab and a new marble cross. Henry Morant and Peter Handcock were court-martialled by the British Army for shooting a German missionary (whom they thought was a spy) and for killing several Boer prisoners. Lieutenants Morant and Handcock strongly argued at their trial that they were acting under higher military orders to take no prisoners whilst fighting the tough Boers of Spelonken, a wild and desolate part of Northern Transvaal. Nevertheless, the two men were found guilty for this serious infringement of British Army discipline and were shot by a firing-squad on 27th. February, 1902. Even today, people still protest that Morant and Handcock were simply "scapecoats of the Empire" whose trial was intended to be a warning to other soldiers who might think of disobeying orders — and that they were executed in order to pacify Germany for the killing of its missionary.

The many necessities carried on the horse's back, along

with its rider, would have made even the strongest animal wilt. A soldier's horse was weighed down with items such as heel-ropes, forage-net, cape, hammer, iron-pegs, blanket, head-rope, horse-brush, horse-sponge, curry-comb, full nosebag, socks, tins of grease, towels, soap, two packs of emergency food-rations, canvas slippers, pair of breeches, two pairs of khaki trousers, clothes-brush, jacket, drill-coat, bootlaces, shoe-case, sponge, flannel shirts, field-cap, 40 rounds of ammunition, riding-pants and an extra pair of boots.

"ROOINEK" (meaning "RED NECK") was an Afrikaans word used in a derogatory manner by the Boers to describe English soldiers; for they soon noticed that these men from a northern climate had pale skins that turned red when exposed to their harsh African sunshine.

"I JOINED A CONTINGENT"

I joined a contingent and went to the war
In search of promotion and pay;
For I fancied the pastime of hunting the Boer
Would suit me at five bob a day.
But, riding along I heard something go *wheeeeee,*
As a bullet passed close to my head;
So I sprang to the ground like a hero so true,
And I lay there pretending to be dead.

(*The officer shouted out, "Forwards! Forwards! so I lay there and said to myself, "Well, he doesn't want me — I'm not a forward, I'm a full-back."*)

So they called me a white-livered cur,
   A Mounted Australian Cur;
   But when all's said and done,
   I'm alive and they're dead,
   And that's what I greatly prefer!
   I'm alive so I'm happy
   With five bob a day
   As a Mounted Australian Cur.

(Yes, a lot of our poor fellows got shot. And what's the good of five bob a day to a dead man? The officer says to me, "You're a white-livered cur." I says, "You're a silly ass. You'll get somebody hurt, letting the Boer shoot at us like this.")

My grandmother taught me a little refrain,
A sort of a nursery rhyme;
"If you fight till you're slain, you cannot fight again,
If you run you can run every time."
And this piece of wisdom I gathered from her,
Which, somehow, I never forgot;
"It will hurt you, no doubt, to be branded a cur,
But it hurts you much more to be shot."

(The officer says, "You'll be shot tomorrow morning by a firing-squad." I says, "No, I won't. I'll go home tonight. I'm tired of the war.")

SELECTED SOURCES

"Official Records of the Australian Military Contingents to the War in South Africa" by Lieutenant-Colonel P. L. Murray (1982)

Boer War Memorial at Donald
(unveiled 15th. October. 1999)

THE GREAT BOER WAR

During the South African War (1899-1902) very few prisoners were taken by the Boers, Anyway, far more dangerous than any human enemy was the dreaded enteric fever which killed thousands of British soldiers. It was indeed a tragic aspect of this war that had water caused more deaths than Boer bullets.

Excellent horsemen, the Boers would move swiftly across the veldt using "hit-and-run" tactics against the British. They did not want prisoners, but they were desperate for clothing and guns; so if a British soldier happened to fall into their hands they would keep his rifle and uniform and then set him free to walk, naked and blushing, back to his camp.

The Boers' most famous prisoner was General Lord Methuen. He was captured at Tweebosch, on 7th. March, 1902, when he and his column were suddenly surrounded by a large body of Boers charging straight at them, five lines deep, and firing from their saddles. It was a hopeless situation for the British.

Surprisingly, Delarey and his men showed kindness and courtesy to the General and towards the 121 wounded and 205 prisoners, returning them all to the British lines — and Delarey even sent his own doctor to escort the wounded.

This amazing act of mercy and charity to prisoners-of-war did much to soften the attitude of the British public towards the Boers and made even the most cynical admit that the Age of Chivalry was not completely over.

Trooper Parker — Front Row.

41071 Trooper
HERBERT. P. PARKER
2ⁿᵈ Scottish Horse (S. Africa)

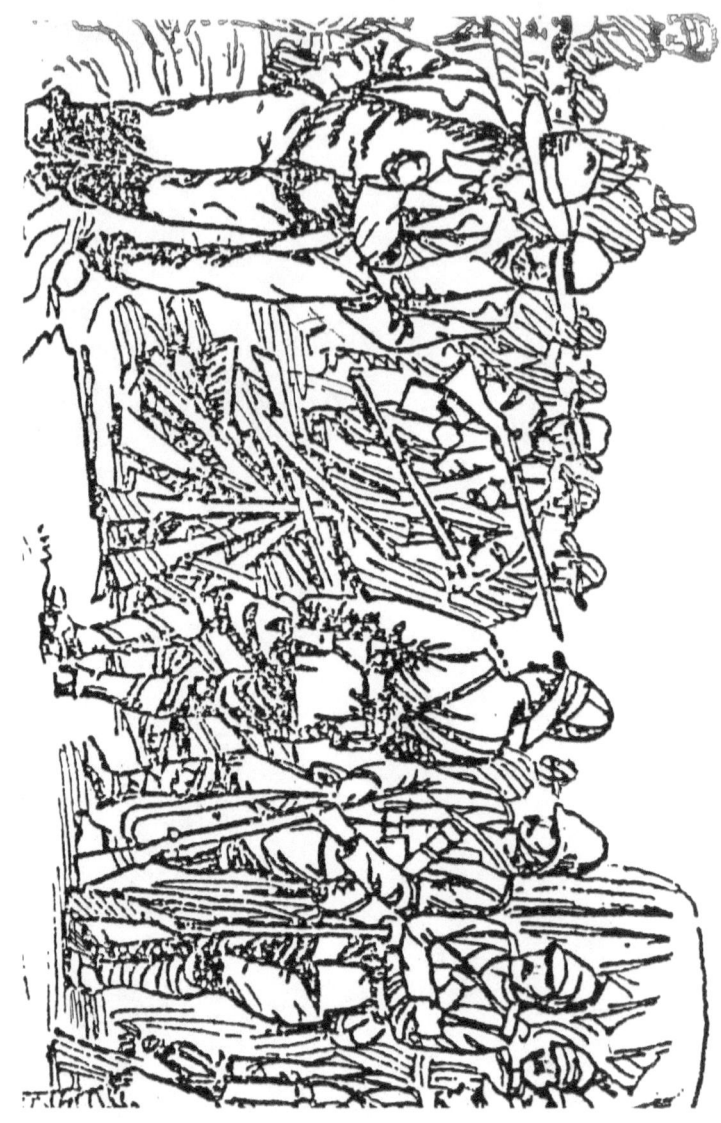

27th. February, 1900. Surrender to the British of General Cronje and his army

SURRENDER OF CRONJE.

Forgotten Heroes

"THE OLD BRIGADE" BY CHARLIE MURRAY (1961)

Tonight the Old Brigade have met, good comrades all are we,
The Boys who many years ago set sail across the sea;
The Boys who wrote a chapter in Australia's history,
And now we **still** go marching on.
When Britain's bugles sounded near the close of '99,
We heard the call and volunteered as soldiers of the line;
We came from farm and city, from sheep — station and gold mine,
And we **still** go marching on.
We sailed to sunny Africa six thousand miles away,
We rode the Veldt together and we climbed the kopjes grey
In full pursuit of sly De Wet, or wily De La Rey,
And we kept **them** marching on.
How often we were hungry as we crossed the rolling Veldt,
When "Bully Beef" and biscuits in half-ration lots were dealt;
But we would not be disheartened and just tightened up our belt,
And we **still** kept marching on.
From battle wounds and fever many gallant comrades died,
Across the Veldt their lonely graves are scattered far and wide;
We honour them in silence and we speak of them with pride,
For their **souls** go marching on.
Our ranks are growing thinner as the long years pass away
But though our joints are stiffer and our locks are silver-grey
The Spirit of the Old Brigade is just as keen today, And we **still**
    go marching on.
GLORY, GLORY TO AUSTRALIA'S OLD BRIGADE,
GLORY, GLORY TO THE SACRIFICES MADE;

"THE OLD BRIGADE" by CHARLIE MURRAY (1961)

GLORY, GLORY TO THE COMRADESHIP DISPLAYED,
AS THEY ALL WENT MARCHING ON.

These verses were written especially for a luncheon given by the Government of Victoria on 9th. October, 1961, in honour of the South African War Veterans whose surviving numbers attended. The above words were sung to the tune of "John Brown's Body". Trooper Clive Dana Colles attended the ceremony.

SOUTH AFRICAN WAR 1899-02

THE BIRTH OF A NATION

The Declaration of War 12th October, 1899

BELFAST
BELMONT
BLOEMFONTEIN TAKEN
COLESBERG
COLENSO
DE WER PURSUIT
DIAMOND HILL
DUNDEE
ELANDSLAAGTE
ELANDS RIVER
GLENCOE
GRASPAN
JOHANNESBURG
KOORN SPRUIT
LADYSMITH
MAJESFONTEIN
MODDER RIVER
PAARDEBERG
PRETORIA TAKEN
RELIEF OF KIMBERLEY
RELIEF OF LADYSMITH
RELIEF OF MAFEKING
RHENOSTER KOP
SPION KOP
SURRENDER OF CRONJE
TALANA HILL
TUGELA HEIGHTS
WILMANS RUST
WEPENER
WITTEBIRGEN
ZAND RIVER
THE SIGNING OF PEACE, 31st, MAY, 1902

THE EXHORTATION
Will you come with me in silent thought, to the Battle Fields of South Africa — to the lonely graves scattered over the veldt-And to the thousands of others who have died of WOUNDS, DISEASE, and SICKNESS since that great ADVENTURE. AMEN.

Forgotten Heroes

MAJOR GENERAL ROBERTS S. BADEN-POWELL

GENERAL OFFICER COMMANDING FRONTIER FORCE

The famous Table Mountain, SOUTH AFRICA.

MAJOR GENERAL ROBERTS S. BADEN-POWELL

The south african war.
Troops storming a 'kopje' or small hill.

www.ingramcontent.com/pod-product-compliance
Lightning Source LLC
Chambersburg PA
CBHW030250010526
44107CB00053B/1654